SolidWorks Electrical 2023
Black Book

By
Gaurav Verma
&
Matt Weber
(CADCAMCAE Works)

ISBN-13 # 978-1-77459-089-8

NOTICE TO THE READER

DEDICATION

To teachers, who make it possible to disseminate knowledge
to enlighten the young and curious minds
of our future generations

To students, who are the future of the world

THANKS

To my friends and colleagues

To my family for their love and support

To Jim Pytel, Columbia Gorge Community College for allowing us to reference his
YouTube channel on basic concepts of electrical technology in this book

Training and Consultant Services

At CADCAMCAE WORKS, we provides effective and affordable one to one online training on various software packages in Computer Aided Design(CAD), Computer Aided Manufacturing(CAM), Computer Aided Engineering (CAE), Computer programming languages(C/C++, Java, .NET, Android, Javascript, HTML and so on). The training is delivered through remote access to your system and voice chat via Internet at any time, any place, and at any pace to individuals, groups, students of colleges/universities, and CAD/CAM/CAE training centers. The main features of this program are:

Training as per your need

Highly experienced Engineers and Technician conduct the classes on the software applications used in the industries. The methodology adopted to teach the software is totally practical based, so that the learner can adapt to the design and development industries in almost no time. The efforts are to make the training process cost effective and time saving while you have the comfort of your time and place, thereby relieving you from the hassles of traveling to training centers or rearranging your time table.

Software Packages for Training

CAD/CAM/CAE: CATIA, Creo Parametric, Creo Direct, SolidWorks, Autodesk Inventor, Solid Edge, UG NX, AutoCAD, AutoCAD LT, EdgeCAM, MasterCAM, SolidCAM, DelCAM, BOBCAM, UG NX Manufacturing, UG Mold Wizard, UG Progressive Die, UG Die Design, SolidWorks Mold, Creo Manufacturing, Creo Expert Machinist, NX Nastran, Hypermesh, SolidWorks Simulation, Autodesk Simulation Mechanical, Creo Simulate, Gambit, ANSYS and many others.

We also provide consultant services for Design and development on the above mentioned software packages

For more information you can mail us at:
cadcamcaeworks@gmail.com

TABLE OF CONTENTS

Chapter 3 : Line Diagram

Chapter 4 : Schematic Drawing

Chapter 5 : Wire Numbering and Customizing

Chapter 6 : Cabinet Layout

Chapter 7 : Practical and Practice

Chapter 8 : Electrical 3D

Annexure I : Basics of Electrical System

Preface

SolidWorks Electrical 2023 is a uniquely designed electrical CAD package from Dassault System. Easy-to-use CAD-embedded electrical schematic and panel designing enable all designers and engineers to design most complex electrical schematics and panels. You can quickly and easily employ engineering techniques to optimize performance while you design, to cut down on costly prototypes, eliminate rework and delays, and save you time and development costs. SolidWorks Electrical provides thousands of symbols and over 500,000 manufactured parts for use in your design hence saving lots of user time for designing rather than drafting.

The **SolidWorks Electrical 2023 Black Book** is, the revised and updated, 9th edition of SolidWorks Electrical Black Book, written to help professionals as well as learners in performing various tedious jobs in Electrical control designing. The book follows the best proven step by step methodology. This book is more concentrated on making you able to use tools at right places. The book starts with basics of Electrical Designing, goes through all the Electrical controls related tools and ends up with practical examples of electrical schematics. Chapters also cover Reports that make you comfortable in creating and editing electrical component reports. There are two annexures added to explain basic concepts of control panel designing. Some of the salient features of this book are:

In-Depth explanation of concepts
Every new topic of this book starts with the explanation of the basic concepts. In this way, the user becomes capable of relating the things with real world.

Topics Covered
Every chapter starts with a list of topics being covered in that chapter. In this way, the user can easy find the topic of his/her interest easily.

Instruction through illustration
The instructions to perform any action are provided by maximum number of illustrations so that the user can perform the actions discussed in the book easily and effectively. There are about 560 illustrations that make the learning process effective.

Tutorial point of view
The book explains the concepts through the tutorial to make the understanding of users firm and long lasting. Each chapter of the book has tutorials that are real world projects.

Project

Projects and exercises are provided to students for practicing.

For Faculty

If you are a faculty member, then you can ask for video tutorials on any of the topic, exercise, tutorial, or concept. As faculty, you can register on our website to get electronic desk copies of our latest books. Faculty resources are available in the **Faculty Member** page of our website (**www.cadcamcaeworks.com**) once you login. Note that faculty registration approval is manual and it may take two days for approval before you can access the faculty website.

Formatting Conventions Used in the Text

All the key terms like name of button, tool, drop-down etc. are kept bold.

Free Resources

Link to the resources used in this book are provided to the users via email. To get the resources, mail us at ***cadcamcaeworks@gmail.com*** or ***info@cadcamcaeworks. com*** with your contact information. With your contact record with us, you will be provided latest updates and informations regarding various technologies. The format to write us e-mail for resources is as follows:

Subject of E-mail as ***Application for resources of .... book***.
Also, give your information like

Name:
Course pursuing/Profession:
Contact Address:
E-mail ID:

For Any query or suggestion

If you have any query or suggestion, please let us know by mailing us on ***cadcamcaeworks@gmail.com*** and ***info@cadcamcaeworks.com***. Your valuable constructive suggestions will be incorporated in our books.

About Authors

The author of this book, Matt Weber, has authored many books on CAD/CAM/CAE available already in market. **SolidWorks Electrical Black Books** are one of the most selling books in SolidWorks Electrical field. The author has hands on experience on many CAD/CAM/CAE packages. If you have any query/doubt during your course, then you can contact the author by writing at cadcamcaeworks@gmail.com

The author of this book, Gaurav Verma, has written and assisted in more than 16 titles in CAD/CAM/CAE which are already available in market. He has authored **AutoCAD Electrical Black Books** which are available in both **English** and **Russian** language. He has also authored books on vocational courses like Automotive Electrician and Civil Electrician. He has provided consultant services to many industries in US, Greece, Canada, and UK.

Chapter 1

Basics of Electrical Drawings

Topics Covered

The major topics covered in this chapter are:

- *Need of Drawings*
- *Electrical Drawings*
- *Common Symbols in Electrical Drawings*
- *Wire and its Types*
- *Labeling*

NEED OF DRAWINGS

In this book, you will learn about electrical wiring and schematics created by using tools in SolidWorks Electrical. Most of the readers of this book will be having prior experience with electrical drawings but there are a few topics that should be revised before we move on to practical on software.

When we work in an electrical industry, we need to have a lot of information ready like the wire type, location of switches, load of every machine, and so on. It is very difficult to remember all these details if you are working on electrical system of a big plant because there might be thousands of wires connecting hundreds of switches and machines. To maintain accuracy in wiring of such big systems, we need electrical drawings. Earlier, electrical drawings were handmade but now, we use printed documentation for these informations. Figure-1 shows an electrical drawing (electronic).

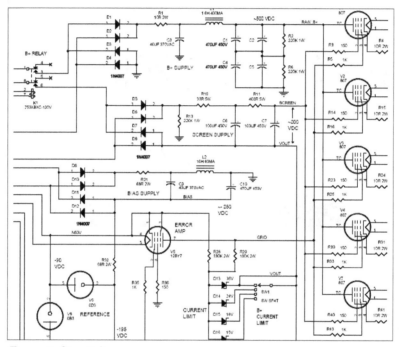

Figure-1. Circuit diagram

ELECTRICAL DRAWINGS

Electrical drawings are the representation of electrical components connected with wiring to perform specific tasks. An electrical drawing can be of a house, industry, or an electrical panel. Electrical drawings can be divided into following categories:

- Circuit diagram
- Wiring diagram
- Wiring schedule
- Block diagram
- Parts list

Circuit Diagram

A circuit diagram shows how the electrical components are connected together. A circuit diagram consists:

- Symbols to represent the components;
- Lines to represent the functional conductors or wires which connect the components together.

A circuit diagram is derived from a block or functional diagram (see Figure-2). It does not generally bear any relationship to the physical shape, size, or layout of the parts. Although, you could wire up an assembly from the information given in it, they are usually intended to show the detail of how an electrical circuit works.

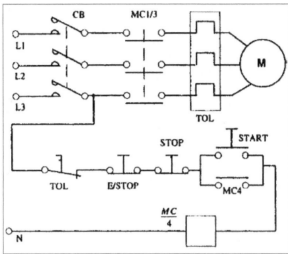

Figure-2. Circuit diagram

Wiring Diagram

A wiring diagram is the drawing which shows complete wiring between the components. We use wiring diagrams when we need to represent:

- Control or signal functions;
- Power supplies and earth connections;
- Termination of unused leads, contacts;
- Interconnection via terminal posts, blocks, plugs, sockets, and lead-throughs.

The wiring diagrams have details, such as the terminal identification numbers which enable us to wire the unit together. Note that internal wiring of components is generally not displayed in the wiring diagrams. Figure-3 shows a wiring diagram.

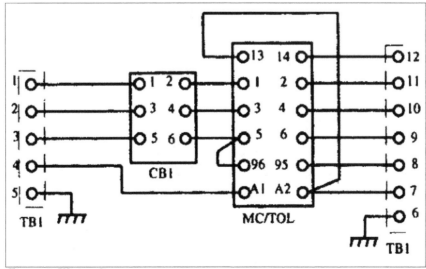

Figure-3. Wiring diagram

Wiring Schedule

A wiring schedule defines the wire reference number, type (size and number of conductors), length, and the amount of insulation stripping required for soldering.

In complex equipment, you may also find a table of interconnections which will give the starting and finishing reference points of each connection as well as other important information such as wire color, identification marking, and so on. Refer to Figure-4.

Schedule:	Motor Control				206-A
Wire No	From	To	Type	Length	Strip Length
01	TB1/1	CB1/1	16/0.2	600 mm	12 mm
02	TB1/2	CB1/3	16/0.2	650 mm	12 mm
03	TB1/3	CB1/5	16/0.2	600 mm	12 mm
04	TB1/4	MC/A1	16/0.2	800 mm	12 mm
05	TB1/5	CtV1	16/0.2	500 mm	12 mm

Figure-4. Wiring Schedule

Block Diagram

The block diagram is a drawing which is used to show and describe the main operating principles of the equipment. The block diagram is usually drawn before creating the circuit diagram.

It will not give the detail of the actual wiring connections or even the smaller components. Figure-5 shows a block diagram.

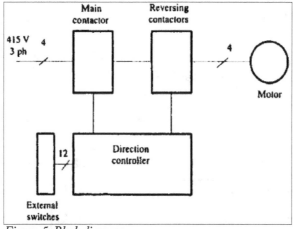

Figure-5. Block diagram

Parts list

Although, Part list is not a drawing in itself but most of the time it is a part of the electrical drawing project; refer to Figure-6. The parts list gives vital information:

- It relates component to circuit drawing reference numbers.
- It is used to locate sand cross refer actual component code numbers to ensure you have the correct parts to commence a wiring job.

PARTS LIST			
REF	**BIN**	**DESCRIPTION**	**CODE**
CB1	A3	KM Circuit Breaker	PKZ 2/ZM-40-8
MC	A4	KM Contactor	DIL 2AM 415/50
TOL	A4	KM Overload Relay	Z 1-63

Figure-6. Parts list

We have discussed various types of electrical drawings and you may have noted that there are various symbols to represent components in these drawings. Following section will discuss some common electrical symbols.

SYMBOLS IN ELECTRICAL DRAWINGS

Symbols used in electrical drawings can be divided into various categories discussed next.

Conductors

There are 12 types of symbols for conductors; refer to Figure-7 and Figure-8. These symbols are discussed next.

1. General symbol, conductor or group of conductors.
2. Temporary connection or jumper.
3. Two conductors, single-line representation.
4. Two conductors, multi-line representation.
5. Single-line representation of n conductors.
6. Twisted conductors. (Twisted pair in this example.)

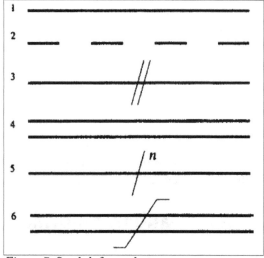

Figure-7. Symbols for conductors

7. General symbol denoting a cable.
8. Multiple conductor (four pair) cable.
9. Crossing conductors – no connection.

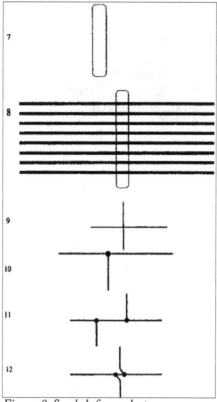

Figure-8. Symbols for conductors

10. Junction of conductors (connected).
11. Double junction of conductors.
12. Alternatively used double junction.

Connectors and terminals

Refer to Figure-9.

13. General symbol, terminal or tag.
14. Link with two easily separable contacts.
15. Link with two bolted contacts.
16. Hinged link, normally open.
17. Plug (male contact).
18. Socket (female contact).
19. Coaxial plug.
20. Coaxial socket.

These symbols are used for contacts with moveable links. The open circle is used to represent easily separable contacts and a solid circle is used for bolted contacts.

Inductors and transformers

Refer to Figure-10.
21. General symbol, coil or winding.
22. Coil with a ferromagnetic core.
23. Transformer symbols.

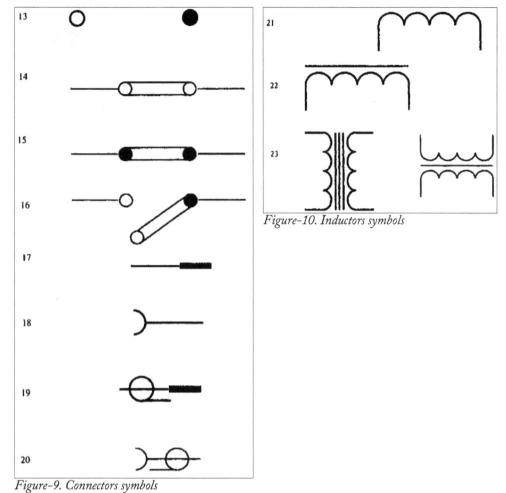

Figure-9. Connectors symbols

Figure-10. Inductors symbols

Resistors

Refer to Figure-11.
24. General symbol.
25. Old symbol sometimes used.
26. Fixed resistor with a fixed tapping.
27. General symbol, variable resistance (potentiometer).
28. Alternative (old).
29. Variable resistor with preset adjustment.
30. Two terminal variable resistance (rheostat).
31. Resistor with positive temperature coefficient (PTC thermistor).
32. Resistor with negative temperature coefficient (NTC thermistor).

Capacitors

Refer to Figure-12.
33. General symbol, capacitor. (Connect either way round.)
34. Polarised capacitor. (Observe polarity when making connection.)
35. Polarized capacitor, electrolytic.
36. Variable capacitor.
37. Preset variable.

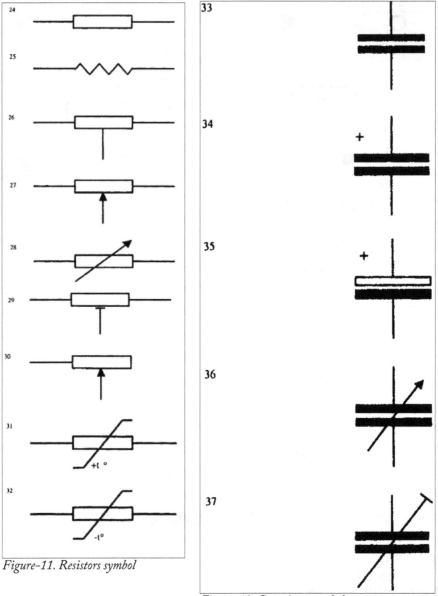

Figure-11. Resistors symbol

Figure-12. Capacitors symbols

Fuses

Refer to Figure-13.

38. General symbol, fuse.

39. Supply side may be indicated by thick line: observe orientation.

40. Alternative symbol (older).

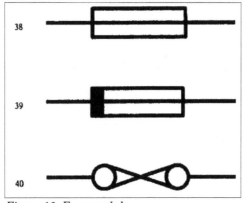

Figure-13. Fuses symbols

Switch contacts

Refer to Figure-14.
41. Break contact (BSI).
42. Alternative break contact version 1 (older).
43. Alternative break contact version 2.
44. Make contact (BSI).
45. Alternative make contact version 1.
46. Alternative make contact version 2.
47. Changeover contacts (BSI).
48. Alternative showing make-before-break.
49. Alternative showing break-before-make.

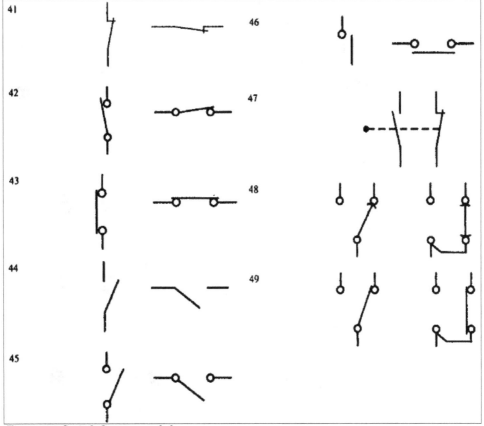

Figure-14. Switch Contact symbols

Switch types

Refer to Figure-15.
50. Push button switch momentary.
51. Push button, push on/push off (latching).
52. Lever switch, two position (on/off).
53. Key-operated switch.
54. Limit (position) switch.

Diodes and rectifiers

Refer to Figure-16.

55. Single diode. (Observe polarity.)
56. Single phase bridge rectifier.
57. Three-phase bridge rectifier arrangement.
58. Thyristor or silicon controlled rectifier (SCR) general symbol.
59. Thyristor – common usage.
60. Triac – a two-way thyristor.

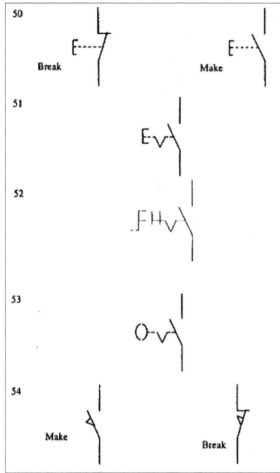

Figure–15. Switch symbols

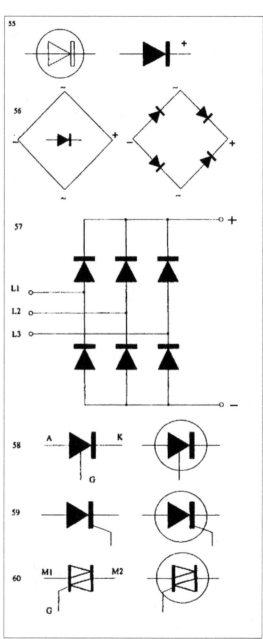

Figure–16. Diode Symbols

Earthing

Refer to Figure-17.

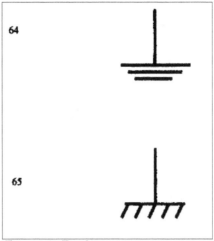

Figure-17. Earthing

Although there are lots of standard symbols in the electrical library, you might still need some user defined symbols for your drawing. You can create these symbols and give map keys/legends for them in the drawing. You will learn about creating symbols later in the book

After learning about various symbols, the next important topic is wire and its specifications.

WIRE AND SPECIFICATIONS

Electrical equipment use a wide variety of wires and cables. It is the responsibility of designer to correctly identify and create the wires in the drawing which are suitable for the current application. The wrong wire types can cause operational problems and could render the unit unsafe. Some factors to be considered for wire selection are:

- The insulation material;
- The size of the conductor;
- Conductor material;
- Solid or stranded and flexible.

Types of Wires

- **Solid or single-stranded wires** are not very flexible and are used where rigid connections are acceptable or preferred usually in high current applications like in power switching contractors. This type of wire can be un-insulated.
- **Stranded wire** is flexible and most interconnections between components are made with it.
- **Braided wire**, also called Screened wire, is an ordinary insulated conductor surrounded by a conductive braiding. In this type of wire, the metal outer is not used to carry current but is normally connected to earth to provide an electrical shield to the internal conductors from outside electromagnetic interference.

Wire specifications

There are several ways to represent wire specifications in electrical drawings. The most common method is to specify the number of strands in the conductor, the diameter of the strands, the cross sectional area of the conductor, and then the insulation type. Refer to Figure-18:

• The 1 means that it is single conductor wire.
• The conductor is 0.6 mm in diameter and is insulated with PVC.
• The conductor has a cross-sectional area nominally of 0.28 mm.

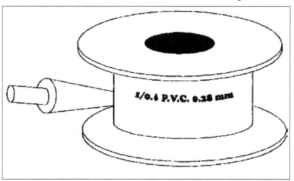

Figure-18. Example of wire specification

Standard Wire Gauge

If you are using solid wire in your drawings then it can be represented in drawing by using the Standard Wire Gauge or SWG system. The SWG number is equivalent to a specific diameter of conductor; refer to Figure-19.

For example;   30 SWG is 0.25 mm diameter.
                  14 SWG is 2 mm in diameter.

The larger the number – the smaller the size of the conductor.

There is also an American Wire Gauge (AWG) which uses the same principle, but the numbers and sizes do not correspond to those of SWG.

SWG table	
SWG No.	**Diameter**
14 swg	2 mm
16 swg	1.63 mm
18 swg	1.22 mm
20 swg	0.91 mm
22 swg	0.75 mm
24 swg	0.56 mm
25 swg	0.5 mm
30 swg	0.25 mm

Figure-19. SWG table

Till this point, we have learned about various schematic symbols and wires. Now, we will learn about labeling of components.

LABELING

Labeling is the marking of components for identifying incoming and outgoing supply; refer to Figure-20.

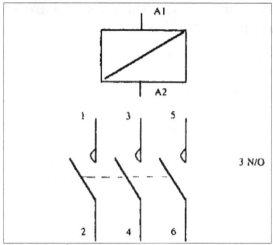
Figure-20. Contacts

Coils are marked alphanumerically, e.g. A1, A2.
  Odd numbers – incoming supply terminal.
  Even numbers – outgoing terminal.

Main contacts are marked with single numbers:
  Odd numbers – incoming supply terminal.
  Next even number – outgoing terminal.

We will find different type of markings while working on the electrical drawings.

Cable Markers

Cable or wire markers are used to identify wires, especially in multi way cables or wiring harnesses. Both ends of the wires are marked with the same numbers to identify start and end of the wires.

Often the cable/wire numbers are same as those of connectors to which they are connected. In any case, the wiring drawing or run-out sheet will give the wire numbers to be used. The markers are placed so that the number is read from the joint as illustrated. This example shows wire number 27; refer to Figure-21.

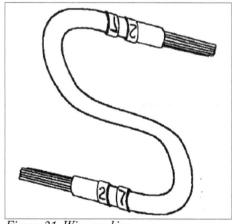
Figure-21. Wire marking

Most of the cables have numbers printed as well as being colored, although you may find some wires/cables colored only.

Some cable markers are wrapped round the wire and are adhesive, while others are like small sleeves which slip over the insulation.

SELF-ASSESSMENT

Q1. Define Electrical Drawings and its types.

Q2. What is the difference between circuit diagram and wiring diagram?

Q3. .... is a type of drawing which defines the wire reference number, type (size and number of conductors), length, and the amount of insulation stripping required for soldering.

Q4. The .... is a drawing which is used to show and describe the main operating principles of the equipment and is usually drawn before the circuit diagram is started.

Q5. Which of the following is symbol for twisted pair conductor?

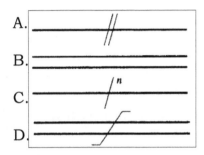

Q6. Which symbol is used to represent easily separable contact?

A. ○

B. ●

Q7. Which of the following wires are not very flexible and are used where rigid connections are accept able or preferred usually in high current applications ?

A. Single-stranded wire
B. Stranded Wire
C. Braided Wire
D. Coaxial Wire

Q8. The larger the SWG number of wire – the smaller the size of the conductor (T/F)

Answer for Self-Assessment:
Ans3. Wiring Schedule, Ans4. Block Diagram, Ans5. D, Ans6.A, Ans7. A, Ans8. T

Chapter 2

Starting with SolidWorks Electrical

Topics Covered

The major topics covered in this chapter are:

- *Installation of SolidWorks Electrical*
- *Starting SolidWorks Electrical*
- *Project Management*
- *Archiving and Un-archiving environment*
- *Configuring Wires, Locations, and Functions*

INSTALLING SOLIDWORKS ELECTRICAL

- If you are installing using the CD/DVD provided by Dassault Systemes then go to the folder containing **setup.exe** file and then right-click on **setup.exe** in the folder. A shortcut menu will be displayed on the screen; refer to Figure-1.

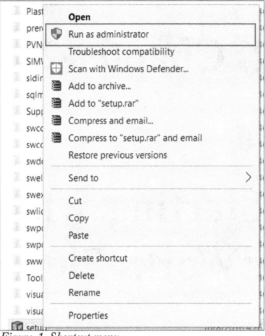

Figure-1. Shortcut menu

- Select the **Run as Administrator** option from the menu being displayed; refer to Figure 1.
- Select the **Yes** button from the dialog box displayed. The **SolidWorks Electrical 2023 Installation Manager** will be displayed. Follow the instructions given in the dialog box. Note that you must have the **Serial Number** of SolidWorks Electrical to install the application. Also, make sure you select the SolidWorks Electrical check box when installing. To know more about installation, double click on the **Read Me** documentation file displayed above the **setup.exe** file.
- If you have downloaded the software from Internet, then you are required to browse in the **SolidWorks Electrical Download** folder in the **Documents** folder. In this folder, open the folder of latest version available and then run **setup.exe**. Rest of the procedure is same.

STARTING SOLIDWORKS ELECTRICAL

- To start SolidWorks Electrical from Start menu, click on the Start button in the Taskbar at the bottom left corner, click on the **All Programs** folder and then on the SolidWorks 2023 folder. In this folder, select the SolidWorks Electrical icon; refer to Figure-2.
- While installing the software, if you have selected the check box to create a desktop icon, then you can double click on that icon to run the software.
- If you have not selected the check box to create the desktop icon but want to create the icon on desktop, then right-click on the **SolidWorks Electrical** icon in the Start menu and select the **Send To-> Desktop (Create icon)** option from the shortcut menu displayed.

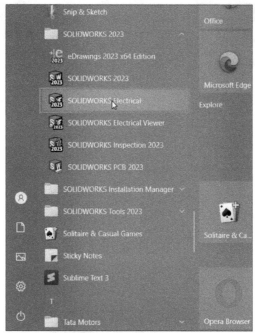

Figure-2. SolidWorks Electrical option in Start menu

After you have performed the above steps, accept the license information. The SolidWorks Electrical 2023 application window will be displayed; refer to Figure-3. Since, this is the first time you are starting SolidWorks Electrical after installation, you are asked to update the library and other data of SolidWorks. Follow the steps given next to update of libraries and SolidWorks data.

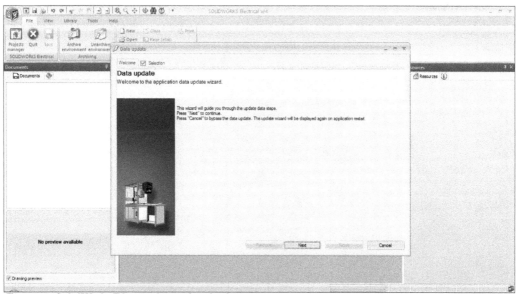

Figure-3. SolidWorks Electrical interface for first time

- Click on the **Next** button from the **Data Update** dialog box. The **Data selection** page of **Data update** dialog box will be displayed; refer to Figure-4.
- The objects in SolidWorks Electrical data that can be selected for update are displayed with their corresponding check boxes. Select the check boxes for updating corresponding objects.
- Note that **Add** option is selected under each category in the page. It means the Project templates will be added in the system library of current. Project template contains unit system, basic files in the form of project, title block, and other related data.

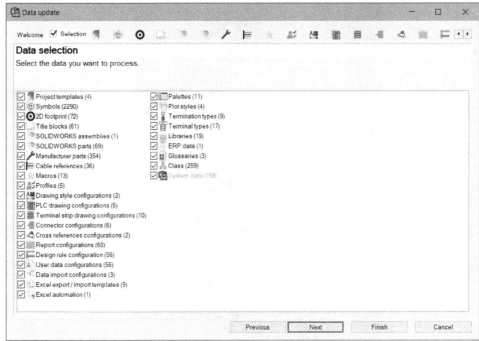

Figure–4. Data Selection page of Data update dialog box

- Click on the **Next** button from the dialog box. Entities in the first selected object category will be displayed; refer to Figure-5.

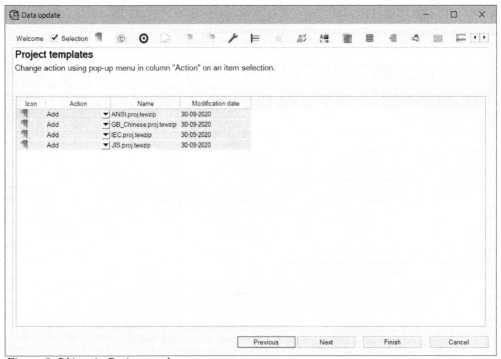

Figure–5. Objects in Project templates category

- Select desired option for each object and click on the **Next** button. A similar page with the objects related to symbols will be displayed.
- Select desired options and keep on clicking next button to add objects in working library. Finally, you will arrive at **Finish** page as shown in Figure-6. Summary of all the items that will be added in library is displayed in this page.
- Click on the **Finish** button from the dialog box. A report will be displayed after update is complete in the **Data update (Report)** page of dialog box.

- Click on the **Finish** button to close the report. The SolidWorks Electrical interface will be displayed with **Electrical Project Management** dialog box as shown in Figure-7.

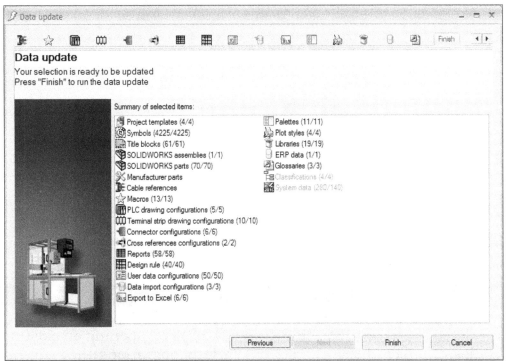

Figure-6. Finish page of Data update dialog box

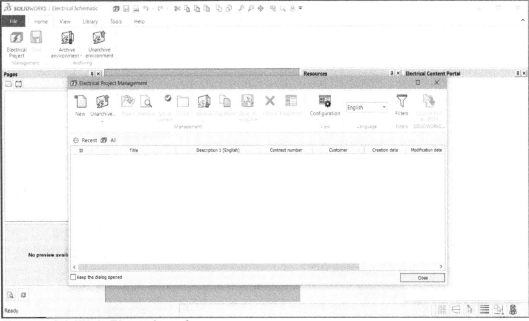

Figure-7. SolidWorks Electrical interface

Note that the first time you start SolidWorks Electrical, you are asked to configure the libraries of electrical components. Like the other electrical CAD packages, SolidWorks Electrical also starts with project setup. Before starting work, you need to create a new project. The options for project setup are available in the **Electrical Project Management** dialog box; refer to Figure-8.

The details of **Electrical Project Management** dialog box are discussed next.

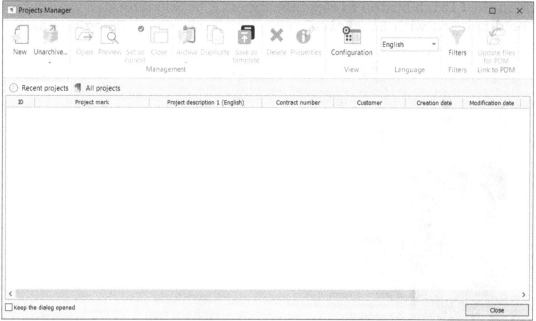

Figure-8. Electrical Project Management dialog box

ELECTRICAL PROJECT MANAGEMENT DIALOG BOX

`Electrical Project Management` dialog box is used to perform all the general operations related to projects like, starting a new project, opening an existing project, creating copies of project files, and so on. Various operations that can be done in `Electrical Project Management` dialog box are discussed next.

Starting a new project

This is the most crucial step for creating electrical drawings in SolidWorks Electrical as all the successive parameters depend on this step. At this step, we decide the electrical standards to be used during the creation of electrical drawings. The steps to start a new project are given next.

*   Click on the **New** button from the **Ribbon** in the `Electrical Project Management` dialog box; refer to Figure-8. The **Create a new project** dialog box will be displayed with various electrical standards that can be used; refer to Figure-9.
*   Select desired option from the drop-down in the dialog box and click on the **OK** button. (We have selected ANSI in our case.)
*   On doing so, the files related to selected standard will be loaded. If you have selected multiple languages during setup, then **Project language** dialog box will be displayed; refer to Figure-10.

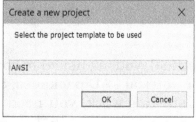

Figure-9. Create a new project dialog box

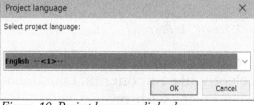

Figure-10. Project language dialog box

*   Select desired language and click on the **OK** button (We have selected **English** in our case). On doing so, the **Project** dialog box will be displayed; refer to Figure-11.

- Specify desired name in the **Name** edit box; refer to Figure-11

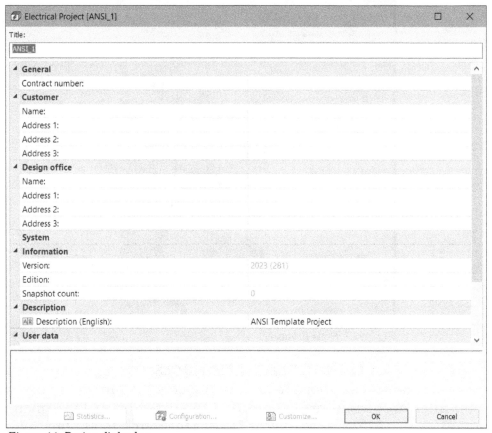

Figure-11. Project dialog box

- One by one, click in the fields of table and specify the data related to customers.
- After specifying the data, click on the **OK** button from the **Project** dialog box. The **SOLIDWORKS Electrical** information box will be displayed notifying you that the database is getting connected for the project; refer to Figure-12.

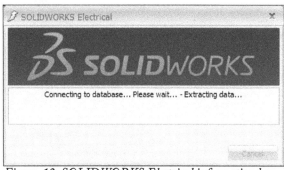

Figure-12. SOLIDWORKS Electrical information box

- Once the process of database linking is completed, the auto-generated document set will be displayed in the **Pages Browser** available in the left of application window; refer to Figure-13. Also, you will be asked whether to open recently opened drawing of the project. Select **No** option to close the drawing or select the **Yes** option to open recent drawing of project.

We will discuss more about **Pages Browser** later in the chapter. Now, we will discuss other options in the **Electrical Project Management dialog box** window.

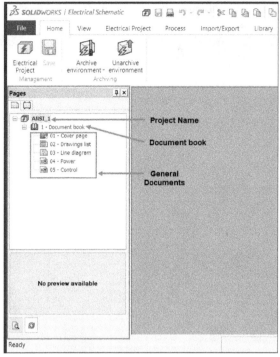

Figure-13. Pages Browser

Un-archiving projects

SolidWorks Electrical provides options to archive projects that are not in use currently. These projects, after archiving, are stored at the server or in local memory. The **Unarchive** option in the **Electrical Project Management** dialog box allows us to unpack those stored projects. The steps to un-archive a project are given next.

* Click on the **Unarchive** tool from the **Electrical Project Management** dialog box. The **Open** dialog box will be displayed and you will be asked to select a SolidWorks Electrical archive file; refer to Figure-14.

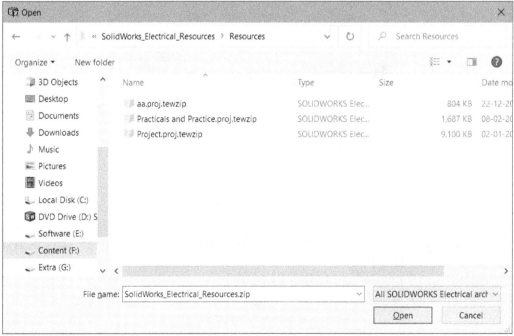

Figure-14. Open dialog box

- Select the archive file that you have earlier saved in your record and click on the **Open** button from the dialog box. The **SolidWorks** information box will be displayed notifying you that the file is being extracted. Once the extraction is complete, the **Project** dialog box will be displayed showing the basic information of the project extracted; refer to Figure-15.
- Click on the **OK** button from the **Project** dialog box to add the extracted project in the **Electrical Project Management** dialog box. The **Merge library elements** dialog box will be displayed asking you whether to update the library with symbols of extracted project or not; refer to Figure-16.

Figure-15. Project dialog box with information of extracted project

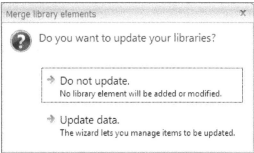

Figure-16. Merge library elements dialog box

- Click on the **Do not update** option from the dialog box and you are asked whether to open the documents of project or not. Open the project to work on it. If you have selected the **Update data** option from the dialog box then **Unarchiving: Electrical Projects** dialog box will be displayed; refer to Figure-17.
- Click on the **Next** buttons from the dialog box and select the check boxes for objects to be imported in the project.
- When you have included all desired components, click on the **Finish** button from the dialog box; refer to Figure-18.

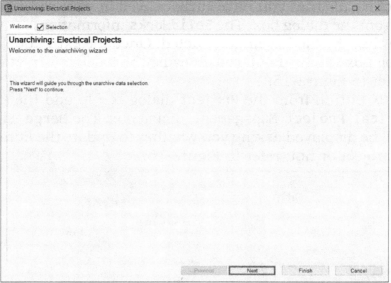

Figure-17. Unarchiving: Electrical Projects dialog box

Figure-18. Finish page of Unarchiving

- Now, open the project to work on it.

Similarly, you can use the **Unarchive and open** tool in **Unarchive** drop-down of **Electrical Project Management** dialog box to unarchive and open the selected project file.

Archiving a project

Archiving of a project is done to save the data related to the project in a compressed folder. This compressed folder can be shared with the peers and customers for further modifications. In previous topic, we have un-archived a project and now we will do the reverse. There are two methods to perform archiving: using **Archive selected projects** and using **Archive Wizard** tools. These methods are discussed next.

Archiving a project using the Archive selected projects tool

- If the project which you want to archive is open then select the project from the **Electrical Project Management** dialog box and click on the **Close** button to close it; refer to Figure-19. On doing so, the project details will be displayed in black color which were bold blue earlier.

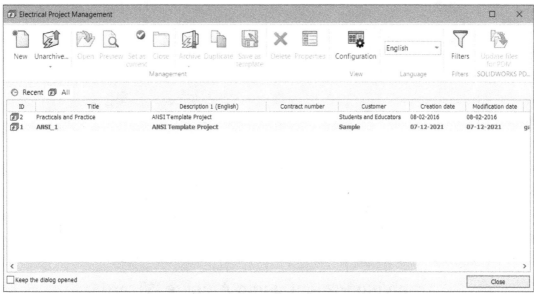

Figure-19. Opened project

- Click on the **Archive selected projects** tool from the **Archive** drop-down in the **Ribbon** of the **Electrical Project Management** dialog box. The **Update reports and terminal strip drawings** dialog box will be displayed; refer to Figure-20.
- Click on the **Update drawings** button if you want to update all the drawing before archiving them. Select the **Continue action without updating** button to archive project without updating drawings. The **Save As** dialog box will be displayed, prompting you to save the archive file; refer to Figure-21.

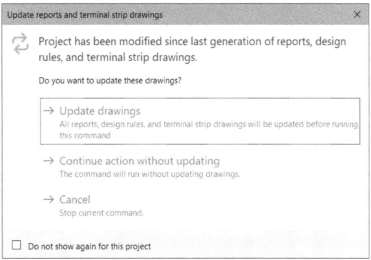

Figure-20. Update reports and terminal strip drawings dialog box

- Specify desired name of the archive and click on the **Save** button to save the file. Once the process of archiving is complete, the **SOLIDWORKS Electrical** dialog box will be displayed asking you whether to open the folder (in which archive is saved) or not.
- Choose the **Yes** option from the information box displayed to open folder for electrical project.

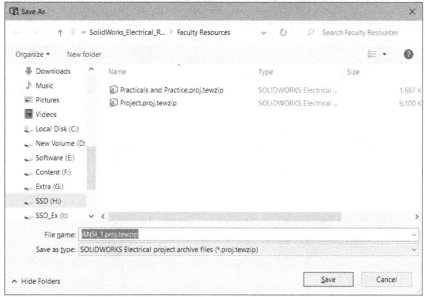

Figure-21. Save As dialog box

Archiving Project using Archive Wizard

You can also use **Archive wizard** tool from the **Archive** drop-down to archive project. The procedure to use this tool is given next.

- After selecting a closed project from the **Electrical Project Management dialog box**, click on the **Archive wizard** tool from the **Archive** drop-down in the **Ribbon** of the dialog box. The **Archive project wizard** dialog box will be displayed; refer to Figure-22.

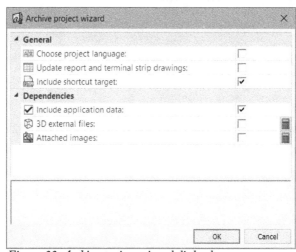

Figure-22. Archive project wizard dialog box

- Select the check boxes for parameters to be archived in project and click on the **OK** button. The **Save As** dialog box will be displayed.
- Specify desired name and click on the **Save** button. The archive file will be saved.

Duplicating Project

Using the **Duplicate** button in the **Electrical Project Management dialog box**, you can create duplicate of a project selected in **Electrical Project Management dialog box**. On choosing this button, you will be asked to specify new name for the duplicate project.

Saving as template

Any of the project you have created earlier can be used as template for successive projects. To make a project as template, follow the steps given next.

*   Click on the **Save as Template** button from the **Electrical Project Management** dialog box after selecting a project. The **Electrical Project** dialog box will be displayed as shown in Figure-23.
*   Specify the name of the template in the **Name** edit box and click on the **OK** button from the dialog box. Once the processing is complete, you can see the new template in the template list used for creating new projects; refer to Figure-24.

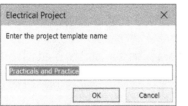

Figure-23. Electrical Project dialog box

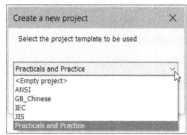

Figure-24. New template created

If you have a long list of projects in the database, use the **Filters** button to filter out the one on which you want to work.

If you work with the peers using the PDM workgroup then click on the **Update files for PDM** button to update file on PDM workgroup so that others can see the latest changes.

Deleting Project

After closing the project to be deleted, select it in the **Electrical Project Management** dialog box and click on the **Delete** button. The **Delete project** dialog box will be displayed; refer to Figure-25.

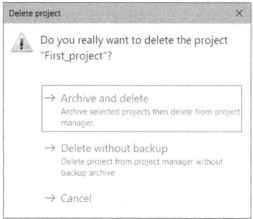

Figure-25. Delete project dialog box

Select the **Archive and delete** button if you want to archive the project and create backup before deleting it. If you want to delete the project without backup copy then select the **Delete without backup** button.

Checking Preview of Project

The **Preview** tool is used to check the files in a closed project. Note that you can only view files by using this tool. You cannot make changes in the files. The procedure to use this tool is given next.

• Select the project you want to check from the **Electrical Project Management** dialog box and click on the **Preview** button from **Management** group in **Ribbon** of **Electrical Project Management** dialog box. The **Electrical Project viewer** window will be displayed; refer to Figure-26.
• Select the drawing you want to check from the bottom list. Preview of the drawing will be displayed in the **Preview** area.
• You can also print and export the selected drawing using the tools in the **Ribbon** of **Project view** window.
• After checking the preview, close the window using **x** button at the top right corner.

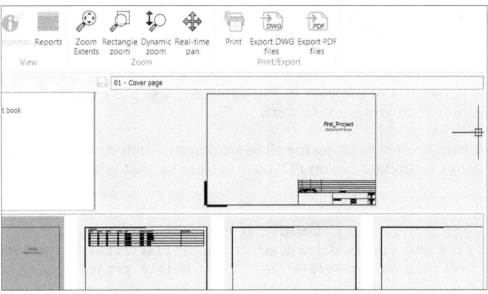

Figure-26. Electrical Project viewer window

Configuring Project

Here, we are not talking about the **Configuration** tool in the **View** panel of the **Electrical Project Management** dialog box. Here, we are talking about configuring a project. The tool to configure a project is available in the **Project** dialog box displayed after selecting **Properties** button from the **Electrical Project Management** dialog box; refer to Figure-27.

• Click on the **Project configuration** button from the **Project** dialog box. The **Project configuration** dialog box will be displayed as shown in Figure-28.

General tab options

• Using the options in the **Project languages** node, you can specify the main language and alternate languages for the project.
• Select desired standard for unit and cable sizes by using the options in the **Standard** node.
• Using the options in the **Date display format** node, you can change the format of date displayed in the drawings of current project.

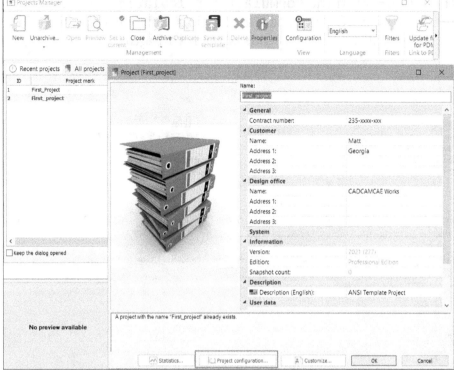

Figure-27. Project configuration button

- Similarly, you can change the revision numbering format by using the **Format** field in the **Revision numbering** node.
- Click in the **Book** field under the **<Default>** node to change the default book of project. Once you have changed the default book, all the new drawings will be automatically added in the selected default book.

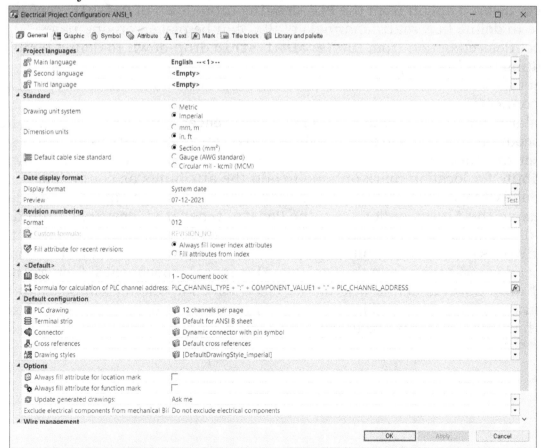

Figure-28. Project configuration dialog box

- Click on the **fx** button in the **Formula for calculation of PLC channel address** field to change the PLC address calculation formula; refer to Figure-29. The **Formula manager** will be displayed as shown in Figure-30.

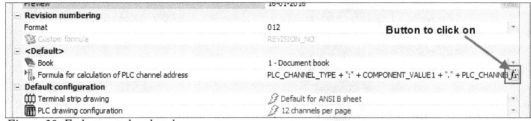

Figure-29. Fx button to be selected

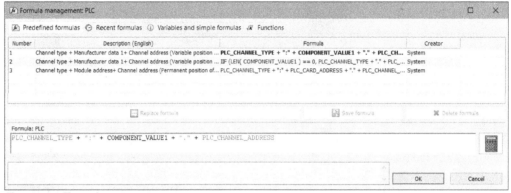

Figure-30. Formula manager

- Select desired formula from the list or add a new formula and select it. You can check various variables in the **Variables and simple formulas** tab of dialog box.
- Click on the **OK** button from the dialog box to apply formula.
- Select desired option from the **PLC drawing** drop-down of **Default configuration** node to define how many channels of PLC can be placed on one page.
- Select desired option from the **Terminal strip** drop-down to define template for creating terminal strip.
- Select desired option from the **Connector** drop-down to define the type of symbol for connector.
- Similarly, set desired options for **Cross references** and **Drawing styles** in their respective drop-downs to define style for cross references and drawing unit system.
- Select the check boxes of Always fill attribute for location mark and Always fill attribute for location mark options to add the attributes as compulsory.
- In SolidWorks Electrical, you can make your drawings update automatically by using the **Always** option from the **Update generated drawings** drop-down in the dialog box; refer to Figure-31. Note that if you now close the project file or SolidWorks Electrical then all the generated drawings will be updated automatically.
- Select desired option from the **Exclude electrical components from mechanical Bill** drop-down to define whether all electrical components will be excluded, no electrical components will be excluded, or electrical components will be excluded but harness will not be excluded.

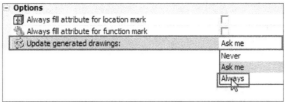

Figure-31. Update generated drawings options

- Similarly, you can set the other options of project.

Graphic tab options

- Click on the **Graphic** tab from the dialog box to display the options related to graphics; refer to Figure-32.
- Set desired size and color for dot placed next to symbol in schematic using the **Symbol** options in the **Connection dots** area of the dialog box.
- Similarly, set desired color and size of dot for wire from the **Connection dots** area.
- You can set the line style parameters in the **Line styles** area of the dialog box.
- Set desired parameters in the **Nodal indicators** area of the dialog box to define how and when nodal indicators will be displayed.

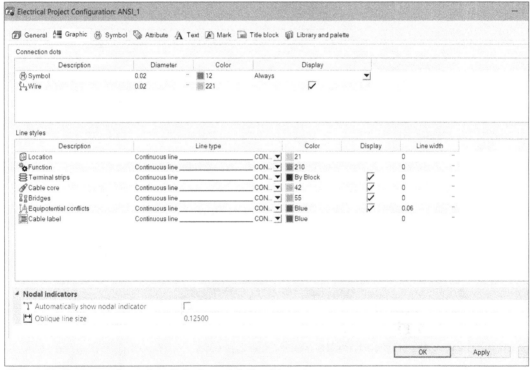

Figure-32. Project configuration dialog box with Graphic options

Symbol tab options

- You can modify the label of cable, wire, location, and equipotential by using the **Select**, **Remove**, or **Modify** buttons displayed at the bottom of each symbol in the dialog box; refer to Figure-33.
- If you wish to modify a tag then click on the **Edit** button for that label in the dialog box. The respective label will open in SolidWorks Electrical in the background.
- Exit the dialog boxes by pressing **ESC** from keyboard. The selected label will be displayed with options to edit; refer to Figure-34.

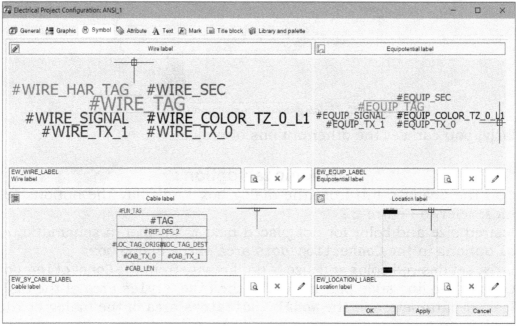

Figure-33. Buttons to modify labels

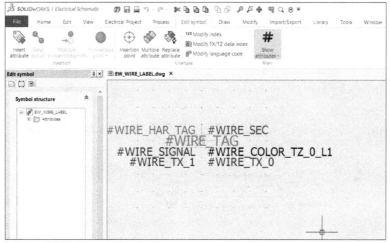

Figure-34. Edit symbol options

Attribute tab options

- Click on the **Attribute** tab from the dialog box to display the options related to attributes; refer to Figure-35. Note that attributes created at project level here can be used by various symbols and title blocks.

Figure-35. Project configuration dialog box with Attribute options

- Click on the **Add custom attribute** button to add new custom attribute for all the other parameters. Set desired parameters.
- Click on the **Add symbol attribute** button to add new attribute for the symbol. The **Attribute Management** dialog box will be displayed; refer to Figure-36. Select desired parameters and click on the **OK** button.

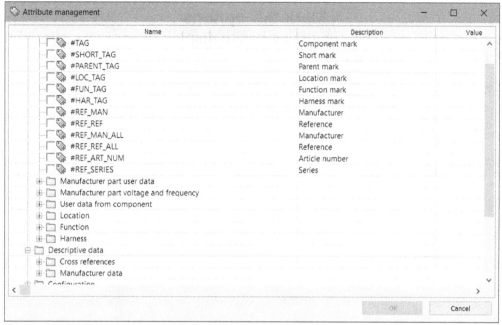

Figure-36. Attribute management dialog box

- Similarly, you can set the other parameters in the **Attribute** tab.

Text tab options

- Click on the **Text** tab from the dialog box to display the options related to fonts; refer to Figure-37.

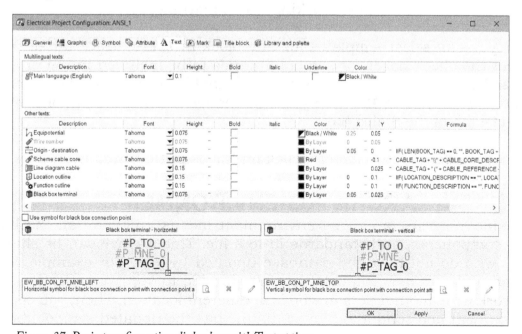

Figure-37. Project configuration dialog box with Text options

- Click on the field under **Font** column and select desired font for the respective object.

• Click on the field under the **Height** column and change the height as per the requirement. Similarly, you can change the other values related to font at the bottom in the dialog box. Note that you can also define formulae for **Scheme cable core** and **Line diagram cable** using the **Fx** option next to their formula field. You will learn more about formulae later in the book.

Mark tab options

• Click on the **Mark** tab from the dialog box to display the options related to various markings; refer to Figure-38.

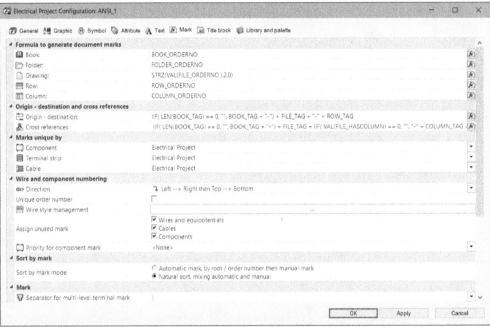

Figure-38. Project configuration dialog box with Mark options

• Using the options in this page, you can change the default marking method for various entities like location marking, cable marking, and so on.

Similarly, you can select templates for title block in **Title block** tab of the dialog box. Select the check boxes for symbol libraries to be included in the project from **Library and palette** tab of the dialog box. Note that if you want to change the symbol palette style from **ANSI** to **IEC** then you can change it using the options in **Library and palette** tab of the dialog box.

• After setting desired parameters, click on the **OK** button from the dialog box.

ARCHIVING ENVIRONMENT

As the name suggests, archiving environment means archiving all the library of symbols, components, and standards in to a file. This archive can be shared with the peers who do not have the database updated by you. For example, you have created around 100 new symbols that are specific to your organization and you want to work with another person who is at different location. Then, you can archive the environment and send it to him so that he has the updated symbol library. The procedure to archive environment is given next.

- Close the **Electrical Project Management** dialog box if it is still open. Click on the **Archive environment** tool from the **Archiving** panel of **File** tab in the **Ribbon**. The **Archiving: Environment** dialog box will be displayed; refer to Figure-39.

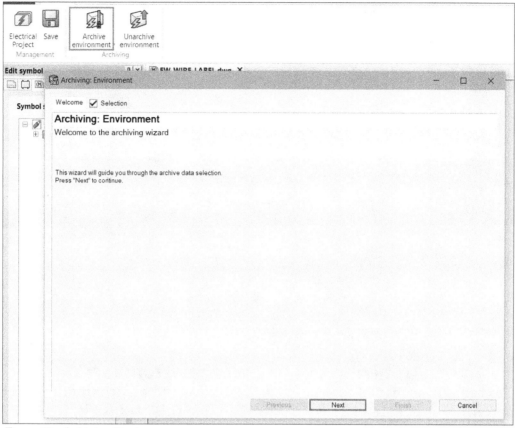

Figure-39. Archiving Environment

- Click on the **Next** button from the dialog box to make selections of the entities. The dialog box will be displayed as shown in Figure-40.
- By default, all the objects are selected for archiving. Click on the **Custom** radio button to select the objects as per your requirement.
- After selecting desired objects, click on the **Next** buttons from the dialog box. At the end, summary page will be displayed.
- Click on the **Finish** button from the dialog box. The **Save As** dialog box will be displayed asking you to save the archive file.
- Specify desired name of the archive and save it at desired location. A dialog box with report will be displayed. Click on the **Finish** button to exit the dialog box displayed. The information box will be displayed asking you whether to open the folder of environment or not. Click on the **Yes** button to check the folder.

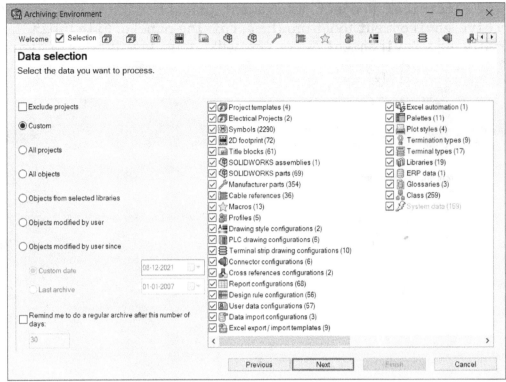

Figure-40. Data Selection page of Archiving Environment

Archiving using External Archiver

The **Run external archiver** tool is used to archive the environment objects using external archiver application. The procedure to use this tool is given next.

- Click on the **Run external archiver** tool from the **Archive environment** drop-down in the **Archiving** panel of **File** tab in the **Ribbon**. The **Environment Archiver** dialog box will be displayed; refer to Figure-41.
- Select desired radio button from the **Archive mode** area to define the scope of objects to be archived.
- Click in the **Output folder** edit box and specify the location where you want to output the archive file.
- After setting desired parameters, click on the **Archive now** button. An information box will be displayed asking you whether to start the process or not.
- Click on the **Yes** button from the box. The application will run in background and once the process is complete, the report will be generated and displayed. If everything goes well, the archive will be generated at specified location.
- Click on the **Close** button from the **Environment Archiver** dialog box to exit the tool.

UNARCHIVE ENVIRONMENT

The **Unarchive environment** tool is available in the **Archiving** panel of **File** tab in the **Ribbon**. The **Unarchive environment** tool makes reversal of **Archive environment** tool. It works in the same way as **Unarchive** tool in the **Electrical Project Management** dialog box.

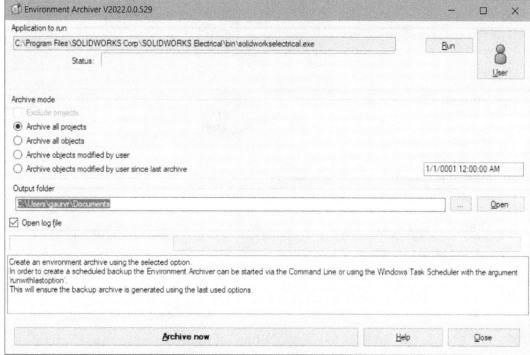

Figure-41. Environment Archiver dialog box

ADDING NEW BOOK/FOLDER IN PROJECT

When we are working on projects that have hundreds of drawings then we generally categorize the files on the basis of their functioning. For example, we are working on a project which has the drawings of electrical distribution of a city. There are n number of companies, houses, and commercial parks; each having their own electrical distribution drawing. So, the drawings of various houses, commercial parks, and companies in the same locality are placed under a book/folder having name of the locality.

Adding New Book

The procedure to add a new book in the project is given next.

- Click on the **New book** tool from the **New** drop-down in the **Electrical Project** panel of the **Electrical Project** tab in the **Ribbon**; refer to Figure-42. The **Book** dialog box will be displayed; refer to Figure-43.

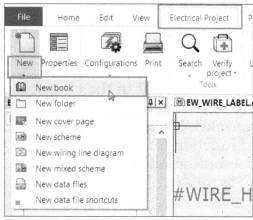

Figure-42. New book tool

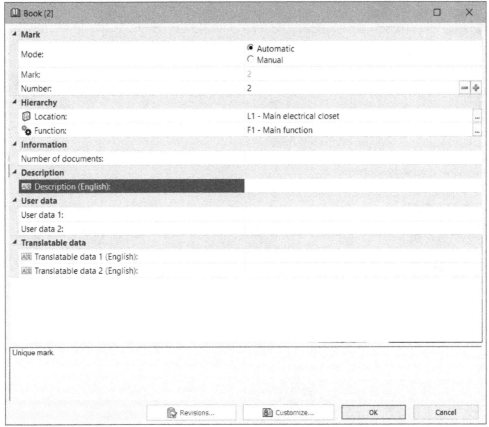

Figure-43. Book dialog box

- A mark number is automatically added to the book which is **2** in our case; refer to **Mark** area in the dialog box shown in Figure-43.
- If you want to manually specify the marking then click on the **Manual** radio button at the top in the dialog box and specify desired mark parameter in the **Mark** edit box; refer to Figure-44. In this figure, we have specified the marking as Street 5-Block 1 which is abbreviated as **St.5-1**.

Figure-44. Manual Marking

- Click on the ellipse button in the field adjacent to **Location** in the **Properties** box of the dialog box. The **Select location** dialog box will be displayed; refer to Figure-45.
- Location is the position of component in the electrical closet. We will learn more about the location later in the chapter.
- Select desired location and click on the **Select** button from the dialog box displayed.
- Similarly, select desired function from the **Select function** dialog box displayed on clicking on the ellipse button adjacent to **Function** in the **Properties** box of the dialog box; refer to Figure-46.

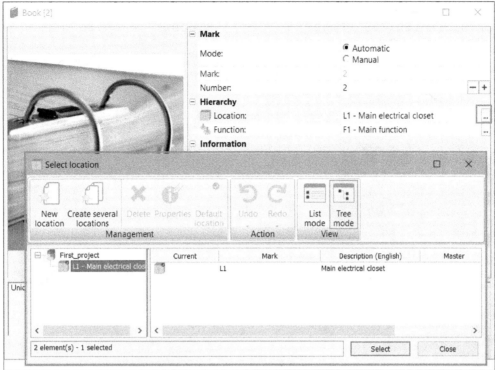

Figure-45. Select location dialog box

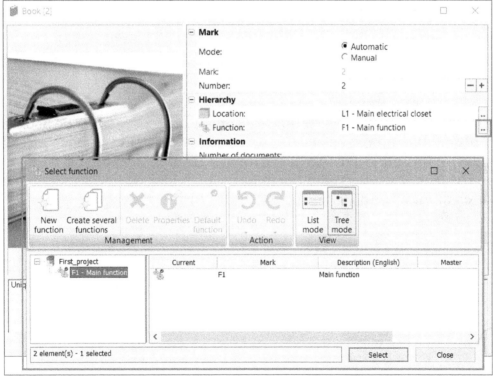

Figure-46. Select function dialog box

- Specify the description and other user information in the **Properties** box and click on the **OK** button from the **Book** dialog box to create the book.

Adding New Folder

The procedure to add a new folder in the project is given next.

- Click on the **New folder** tool from the **New** drop-down in the **Electrical Project** panel of the **Electrical Project** tab in the **Ribbon**. The **Folder** dialog box will be displayed; refer to Figure-47.

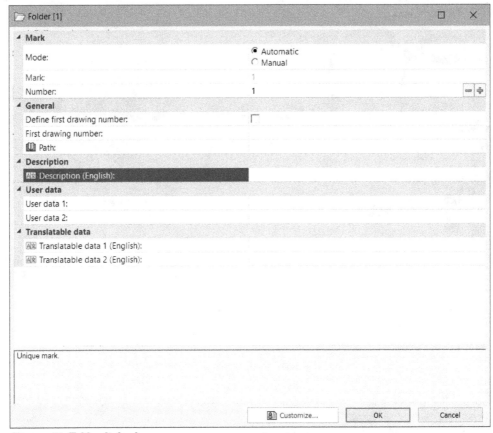

Figure-47. Folder dialog box

- Most of the options in this dialog box are same as discussed for the **Book** dialog box.
- In the **Folder** dialog box, select the **Define first drawing number** check box and specify desired number for the first drawing in the folder if you want to.
- Click on the **OK** button to create the folder. The folder will be added in the selected book; refer to Figure-48.

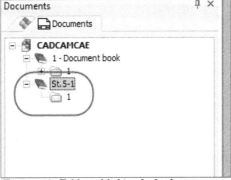

Figure-48. Folder added in the book

ADDING NEW DRAWING IN PROJECT

Cover page, Line diagram, Schema, and Mixed Scheme are collectively called drawings in SolidWorks Electrical. The tools to add these drawings are also available in the **New** drop-down in the **Electrical Project** panel of the **Electrical Project** tab in the **Ribbon**; refer to Figure-49. The procedure to start any of these drawings is similar.

Here, we will discuss the procedure to add a new schematic drawing (schema). You can add the other type of drawings in the same way.

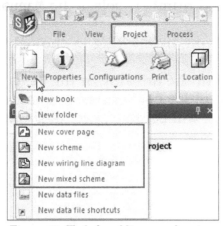

Figure-49. Tools for adding new drawing

Adding Schema

The procedure to add a schematic drawing in the project is given next.

*   Click on the **New scheme** tool from the **New** drop-down in the **Electrical Project** panel of the **Electrical Project** tab in the **Ribbon**. The new scheme will be added in the project.
*   Right-click on the newly added drawing from the **Pages Browser** and select the **Properties** option. The **Drawing** dialog box will be displayed as shown in Figure-50.

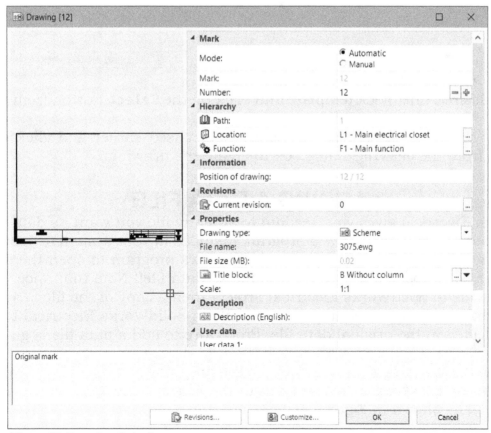

Figure-50. Drawing dialog box

- Mark number is automatically assigned to the drawing since the **Automatic** radio button is selected in the **Mark** area of the dialog box. In our case, the mark number is **6**. To give a user-defined number, select the **Manual** radio button and specify desired mark number.
- Set desired scale for drawing by clicking in the field corresponding to **Scale** in the dialog box.
- Enter the description about the drawing in the **Description (English)** field.
- Click on the Browse button for field corresponding to **Title block:**. The **Title block selector** dialog box will be displayed as shown in Figure-51.

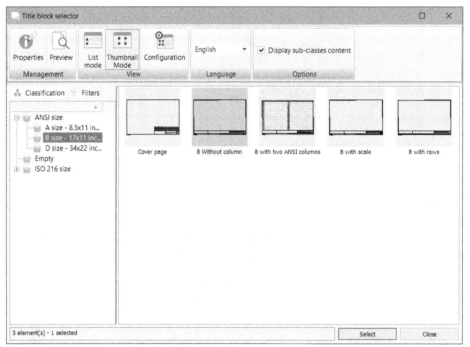

Figure-51. Title block selector dialog box

- Select desired title block template and click on the **Select** button from the dialog box.
- Specify the other desired parameters as discussed earlier and click on the **OK** button from the **Drawing** dialog box to create the drawing.

ADDING A DATA FILE

SolidWorks Electrical gives you freedom to add any file you want as data file in the project. Note that you must have a program installed in your system to open the data file because SolidWorks uses the default Windows program to open the data file. I know some of the people will add movie file as data file!! Note that once you add a file as data file to SolidWorks Electrical Project then a copy of the file is stored with the project files. So if you change the data file in SolidWorks Electrical then it will not be reflected in the original data file. Procedure to add a data file is given next.

- Click on the **New data files** tool from the **New** drop-down in the **Electrical Project** panel of the **Electrical Project** tab in the **Ribbon**. The **Open** dialog box will be displayed as shown in Figure-52.

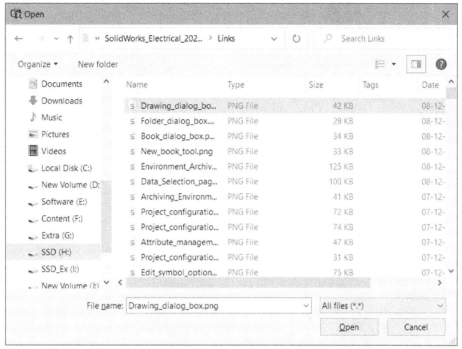

Figure-52. Open dialog box

- Select desired file and click on the **Open** button from the dialog box. The selected file will be added in the current book of the project.

ADDING DATA FILE SHORTCUT

Okay! we have added a data file in SolidWorks project then why do we need a data file shortcut. The answer is: When you add a data file shortcut then on changing the file in SolidWorks Electrical, the original file also changes. This makes the process more streamlined. The procedure to add data file shortcut is given next.

- Click on the **New data file shortcuts** tool from the **New** drop-down in the **Electrical** Project panel of the **Electri**cal Project tab in the **Ribbon**. The **Open** dialog box will be displayed as shown earlier in Figure-52.
- Double-click on the file whose shortcut is to be added in the project. A shortcut file will be added in the project.

CONFIGURING WIRES

Wires are the life-line of any circuit. It is important to configure wires before using them. The procedure to configure wires is given next.

- Click on the **Wire style** option from the **Configurations** drop-down in the **Electrical Project** panel of the **Electrical Project** tab in **Ribbon**; refer to Figure-53. The **Wire style manager** will be displayed; refer to Figure-54.

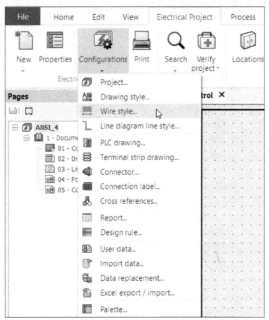

Figure-53. Wire styles option

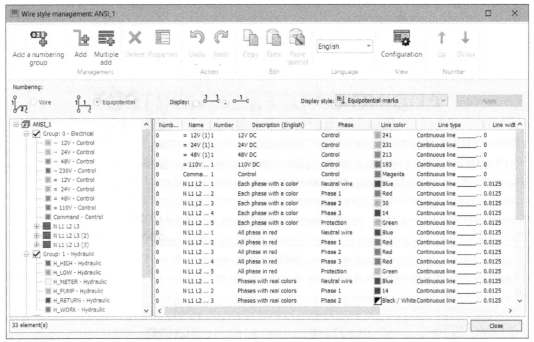

Figure-54. Wire style management

- To modify the wire style, double-click on it in the table. The **Wire style properties** dialog box will be displayed; refer to Figure-55.
- Click in desired field and change the parameters as per your requirement.
- Click on the **OK** button to set the wire style properties.

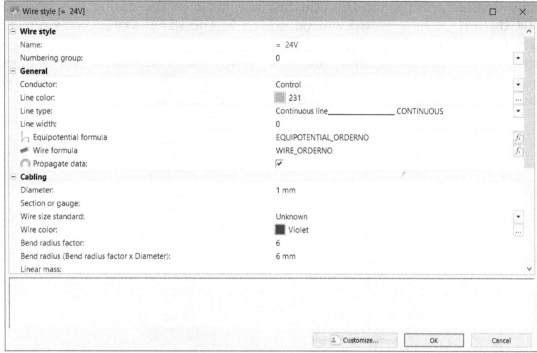

Figure-55. Wire styles properties dialog box

Adding a Numbering Group

A numbering group is used to categorize wires as per their usage. In SolidWorks Electrical, wires with equipotential counter are placed in one group. The procedure to create a numbering group is given next.

- Click on the **Add a numbering group** tool from the **Wire style management** dialog box; refer to Figure-56. The **New numbering group** dialog box will be displayed as shown in Figure-57.

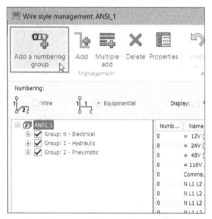

Figure-56. Add a numbering group button

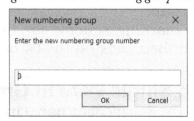

Figure-57. New numbering group dialog box

- Specify desired number in the edit box available in the dialog box and click on the **OK** button. A new group will be added in the **Wire style manager**; refer to Figure-58.
- Right-click on the name of newly created group. A shortcut menu will be displayed.
- Select the **Properties** option from the shortcut menu; refer to Figure-59. The **Numbering group** dialog box will be displayed; refer to Figure-60.

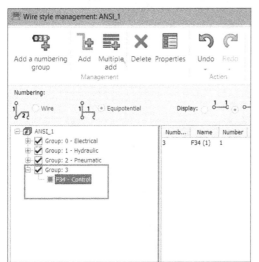

Figure-58. New wiring group

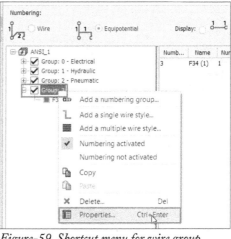

Figure-59. Shortcut menu for wire group

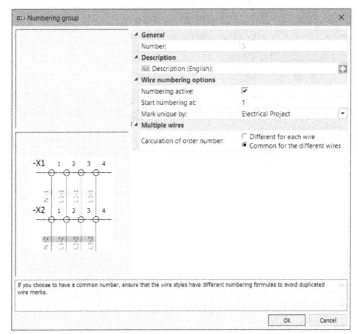

Figure-60. Numbering group dialog box

- Click in the field adjacent to **Description (English)** in the table and specify desired description for the group.
- Specify the numbering and marking scheme for the wires by using the other options in the dialog box and then click on the **OK** button from the dialog box.

Adding Single Wire in Group

- Click on the **Add** button from the **Management** panel in the **Wire style management**. A new wire will be added in the selected group.
- Double-click on the wire in the table. **Wire style** dialog box will be displayed as discussed earlier. Change the properties as discussed earlier.

Adding Multiple Wire in Group

• Click on the **Multiple add** button from the **Management** panel in the **Wire style manager**. A set of multiple wires will be added in the group; refer to Figure-61.

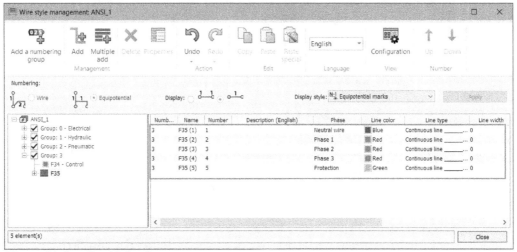

Figure-61. Set of multiple wires

• Expand the node to check individual wires. Double-click on each wire to change its properties.
• Use the **Up** and **Down** buttons in **Number** panel of the **Wire style management** dialog box to change the position and numbering of wire in the dialog box.
• To delete any wire style, select it and press **DELETE** button from the keyboard or click on the **Delete** button from dialog box.

CONFIGURING DRAWING STYLES

Drawing styles are used to define layers, line styles, text styles, and leader styles used in the drawing for representing various objects. The **Drawing styles** tool in the **Configurations** drop-down is used to create and manage drawing styles. The procedure to configure drawing styles is given next.

• Click on the **Drawing style** tool from the **Configurations** drop-down in the **Electrical Project** panel of **Electrical Project** tab in the **Ribbon**. The **Drawing styles management** dialog box will be displayed; refer to Figure-62.

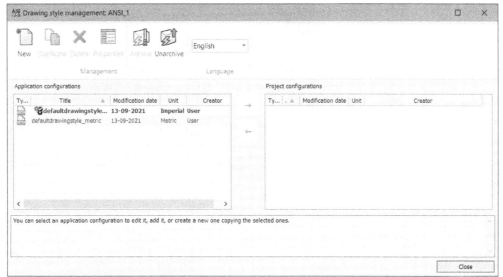

Figure-62. Drawing styles management

- Click on the **New** button from the **Ribbon** in the **Drawing style management** dialog box to add a new drawing style. The **New drawing style** dialog box will be displayed; refer to Figure-63.

Figure-63. New drawing style dialog box

- Specify desired name of drawing style and click on the **OK** button. The new style will be created and the **Edit drawing style** dialog box will be displayed; refer to Figure-64. There are various tabs in the dialog box like **Layer**, **Linetypes**, **Text style**, and so on.
- Create and edit various objects of drawing style using the options in the dialog box and then click on the **OK** button. New style will be displayed in the **Drawing styles manager**.

Figure-64. Edit drawing style dialog box

- If you want to delete a drawing style then select it from the manager and click on the **Delete** button from the **Management (Application)** panel of **Ribbon** in the **Manager**. A warning box will be displayed. Select **Yes** button to delete the configuration.
- You can use the other tools of **Drawing styles manager** as discussed earlier.
- Close the manager by using **x** button at the top right corner.

Note that there are two types of drawing style configurations; application configurations and project configurations. The configurations created for project are specific for current project while the application configurations are stored for use in multiple projects.

We will discuss about the other configurations in their related chapters, later in the book. Now, it's time to understand the concept of location and function.

LOCATIONS

Locations are used to group the components on the basis of their locations in circuit, panel, or placement in floor plan. For example, you have a common panel for three storey building. Then you can define the locations as L0 for base floor, L1 for first floor, and so on. Locations help to identify the components. The procedure to create location codes is given next.

*   Click on the **Locations** tool from the **Management** panel in the **Electrical Project** tab of the **Ribbon**. The **Location management** dialog box will be displayed; refer to Figure-65.

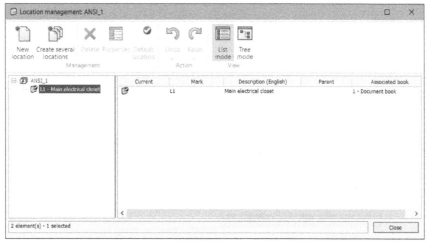

Figure-65. Location management dialog box

*   **L1** is available in the **Location management** dialog box by default. To add more locations, click on the **New location** tool from the **Management** panel in the **Location management** dialog box. The **Location properties** dialog box will be displayed as shown in Figure-66.

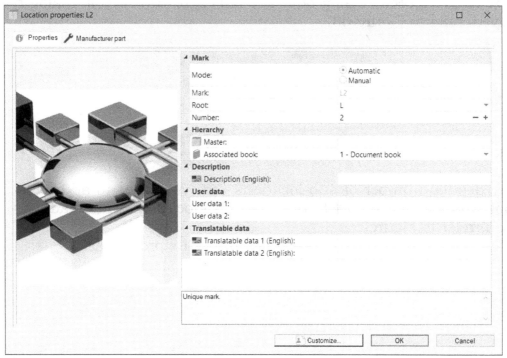

Figure-66. Location properties dialog box

- Click in the **Root** edit box and specify desired keyword for your easy identification like, **Area**.
- Set the identification number by using the **Number** spinner/edit box.
- Click in the field adjacent to **Description (English)** in the table and specify description about the location. Similarly, set the other user data; refer to Figure-67.
- Click on the **Manufacturer part** tab of the dialog box to assign a manufacturer part to current location. Generally a box or cabinet is associated with location. You will learn more about manufacturer parts later in this book.
- After setting the data, click on the **OK** button from the dialog box. The new location will be added in the **Locations management** dialog box.

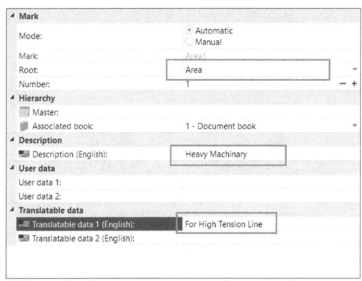

Figure-67. Data specified in Location dialog box

Creating Multiple Sub-locations

- Click on the **Create several locations** tool from **Management** panel in the **Locations management** dialog box. The **Multiple insertion** dialog box will be displayed; refer to Figure-68.

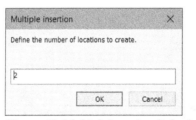

Figure-68. Multiple insertion dialog box

- Specify desired number of sub-locations that you want to add in the selected location and click on the **OK** button. The sub-locations will be added under the location; refer to Figure-69.
- Double-click on the location in table to change its properties.

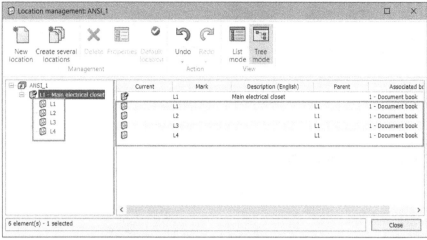

Figure-69. Sub-locations added

FUNCTIONS

Functions are used to identify the components on the basic of their collective function. For example, there are 10 components that are used to control the motors then these components can be put under the function named control. The procedure to add functions is given next.

- Click on the **Functions** tool from the **Management** panel in the **Electrical Project** tab of the **Ribbon**. The **Function management** dialog box will be displayed as shown in Figure-70.

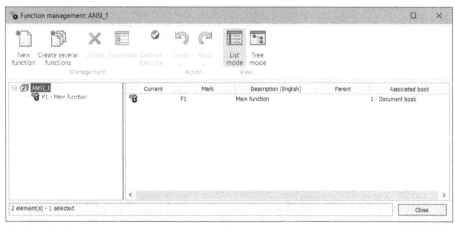

Figure-70. Function management dialog box

- Click on the **New function** button from the **Management** panel in the **Function management** dialog box. The **Function** dialog box will be displayed; refer to Figure-71.
- Specify the parameters as discussed earlier for **Location** dialog box.
- Click on the **OK** button from the dialog box. The function will be added in the **Functions management** dialog box.

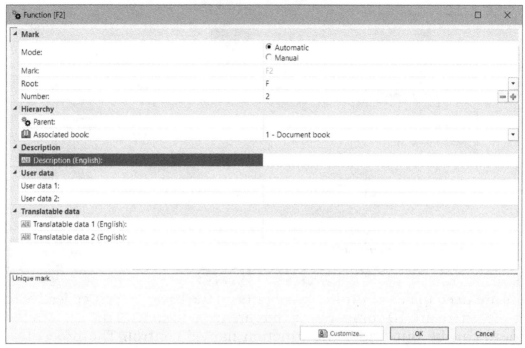
Figure-71. Function dialog box

Creating Multiple Sub-functions

- Click on the **Create several functions** button from the **Management** panel in the **Functions manager**. The **Multiple insertion** dialog box will be displayed; refer to Figure-72.

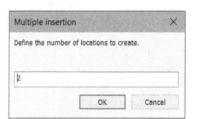
Figure-72. Multiple insertion dialog box

- Specify the number of sub-functions that you want to add in the edit box and click on the **OK** button from the dialog box. The sub-functions will be added in the selected function; refer to Figure-73.

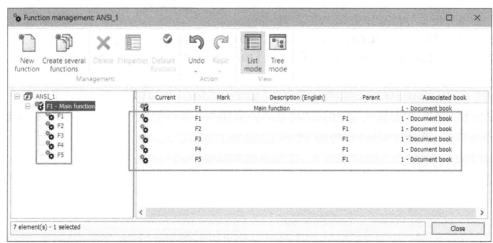
Figure-73. Sub-functions added

- Double-click on the functions to change their properties.

SEARCH TOOLS

There are various tools available in the **Search** drop-down of **Tools** panel in **Electrical Project** tab of **Ribbon** to search for components, wires, locations, and so on; refer to Figure-74. These tools are used to check the objects in projects by using specified parameters. These tools are discussed next.

Figure-74. Search drop-down

Searching by Object Type

The **Search** tool is used to search objects by their object types. The procedure to use this tool is given next.

- Click on the **Search** tool from the **Search** drop-down in the **Tools** panel of **Electrical Project** tab in the **Ribbon**. The **Search in project** dialog box will be displayed; refer to Figure-75.

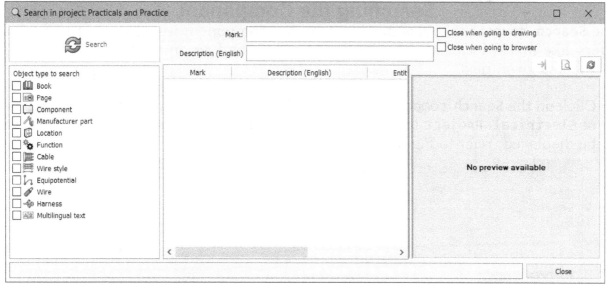

Figure-75. Search in project dialog box

- Select check boxes for the object types to be searched in the project. For example if you are looking for a specific wire in project then select the **Wire** check box from left section of the dialog box.
- Specify the name of object in **Mark** edit box and description of the object in **Description** edit box of the dialog box.
- Select the **Close when going to drawing** and **Close when going to browser** check boxes to close the dialog box when you are looking at the object in drawing and component browser, respectively
- Click on the **Search** button to look for the object based on specified parameters. The list of matching components will be displayed; refer to Figure-76.

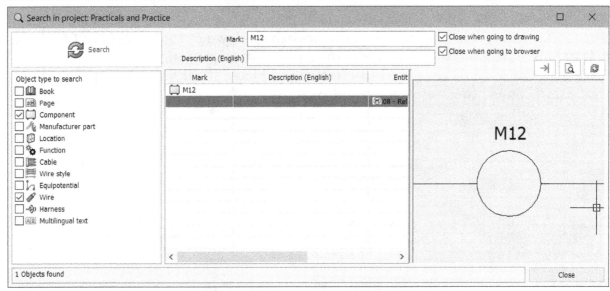

Figure-76. List of searched objects

- Select the object you want to display in drawing from the search list in dialog box and click on the **Go to drawing** ⇥ button to check the object in drawing.
- Select the object that you want to display in component browser from the search list in dialog box and click on the **Go to browser** ⊡ button.
- Click on the **Close** button to exit the dialog box.

Searching the Component

The **Search component** tool is used to search a component in the project by various filters like mark, location, function, description, and so on. The procedure to use this tool is given next.

- Click on the **Search component** tool from the **Search** drop-down in the **Tools** panel of **Electrical Project** tab in the **Ribbon**. The **Search component** dialog box will be displayed; refer to Figure-77.

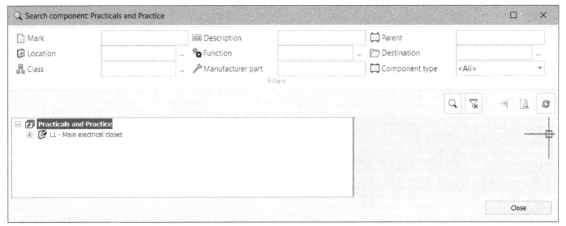

Figure-77. Search component dialog box

- Set desired filter parameters in the dialog box and search the component as discussed earlier.
- The other options of the dialog box have been discussed earlier. Click on the **Close** button to exit the dialog box.

Similarly, you can use **Search page** tool to search for desired drawing in the book.

VERIFYING PROJECT

The **Verify project** tool is used to check integrity of project and fix various common errors of the project. The procedure to use this tool is given next.

- Click on the **Verify project** tool from the **Verify project** drop-down in the **Tools** panel of **Electrical Project** tab in the **Ribbon**. The **Verify project** dialog box will be displayed; refer to Figure-78.

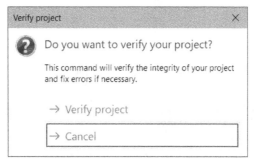

Figure-78. Verify project dialog box

- Click on the **Verify project** button from the dialog box. The **Verify project** information box will be displayed. Once the process is complete, list of errors will be displayed; refer to Figure-79.

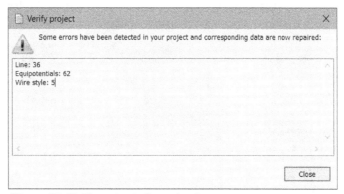

Figure-79. Verify project information box

- Click on the **Close** button to exit the dialog box. The **SOLIDWORKS Electrical** information box will be displayed asking you whether you want to see the list of verifications performed on project; refer to Figure-80.

*Figure-80. SOLIDWORKS Electrical
information box*

- Click on the **Yes** button to check the verification list. The project verification list will be displayed; refer to Figure-81.

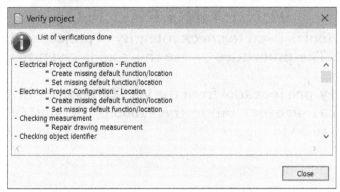

Figure-81. Verification list

- Click on the **Close** button to exit the dialog box.

Purging Project Drawings

The **Purge project drawings** tool is used to delete unused block tables, layer tables, and so on. The procedure to use this tool is given next.

- Click on the **Purge project drawings** tool from the **Verify project** drop-down in the **Ribbon**. The information box will be displayed asking you to confirm whether delete data like text styles, blocks styles, and so on.
- Click on the **OK** button to purge drawings. The **Purge active project drawings** dialog box will be displayed; refer to Figure-82.

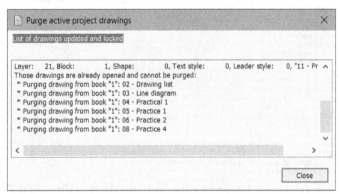

Figure-82. Purge active project drawings dialog box

- After checking the information, click on the **Close** button to exit.

Similarly, you can use **Fix location manufacturers parts** tool to automatically fix errors related to manufacturer parts used in drawings.

Chapter 3

Line Diagram

Topics Covered

The major topics covered in this chapter are:

- *Creating Line diagrams*
- *Inserting Symbols and Manufacturer parts*
- *Connection labels*
- *Drawing cables*
- *Origin - Destination Arrows*
- *Function outline and Location outline*
- *Detailed Cabling*

INTRODUCTION

Line diagrams are used to represent the complete cabling with the help of single lines and components. In case of line diagrams, we do not need to insert detailed schematic diagrams. We insert only simplified representations of components to displayed the cabling arrangement.

CREATING WIRING LINE DIAGRAM

- Click on the **New wiring line diagram** tool from the **New** drop-down in the **Electrical Project** panel of the **Electrical Project** tab in the **Ribbon**. A new line diagram file will be added in the project.
- Right-click on the newly added drawing and select **Properties** option from the shortcut menu. The **Drawing** dialog box will be displayed with the options related to line diagram; refer to Figure-1. You can display the same dialog box by selecting the newly added drawing from **Documents Browser** and press **CTRL** with **ENTER** from keyboard.

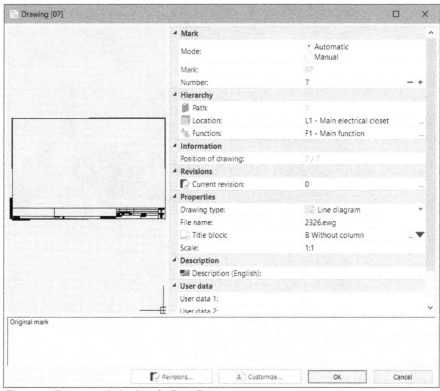

Figure-1. Drawing dialog box for Line Diagram

- Set the drawing number in the **Number** edit box or select **Manual** radio button and specify desired identifier.
- Click on the **Browse** button for the **Location** field and set the location of drawing from the **Select location** dialog box displayed; refer to Figure-2.
- Click in the **Function** field and select desired function for the drawing from the **Select function** dialog box; refer to Figure-3.
- Click in the **Title block:** field and set the title block for the drawing from the drop-down.
- Click in the **Description (English)** field and specify the description about the drawing.
- Click on the **OK** button from the **Drawing** dialog box. The drawing will be added in the project. Since, we have changed location and function of the file so, the **Change**

drawing location dialog box will be displayed asking you if you want to change location of components with drawing or you want to change drawing location only; refer to Figure-4. Select desired option from the dialog box. Similarly, set desired option from the **Change drawing function** dialog box.

Figure-2. Select location dialog box

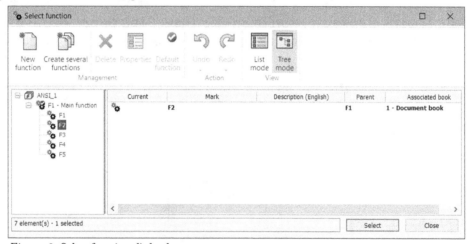

Figure-3. Select function dialog box

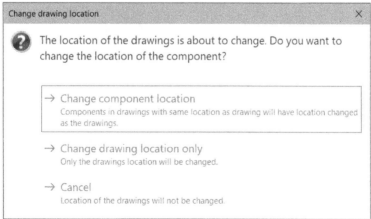

Figure-4. Change drawing location dialog box

- Double-click on the newly added line diagram from the **Pages browser** available at the left of the window. The line diagram will be opened if not open already; refer to Figure-5.

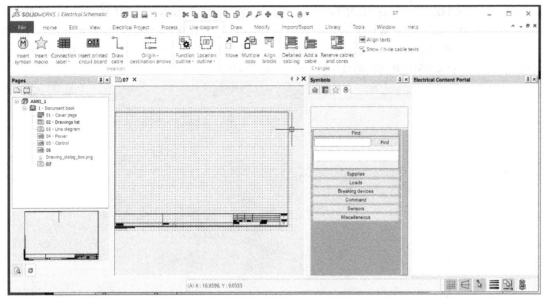

Figure-5. Line diagram opened

INSERTING SYMBOLS

- Click on the **Insert symbol** button from the **Insertion** panel in the **Line diagram** tab of the **Ribbon**. The **Symbol selector** dialog box will be displayed as shown in Figure-6.

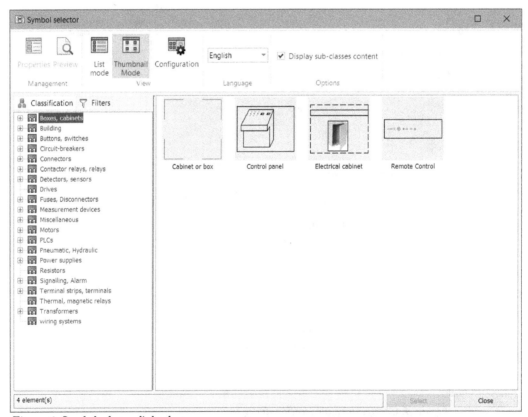

Figure-6. Symbol selector dialog box

- Expand various categories to check the list of components available. After selecting a category, the related components will be displayed on the right in the dialog box. Select desired component from the dialog box and click on the **Select** button. Note that you can apply filters to the list by setting desired parameters in the **Filters** tab of the dialog box. On selecting the component, it will get attached to cursor and the **Symbol insertion** options will be displayed in the **Command Browser**; refer to Figure-7.

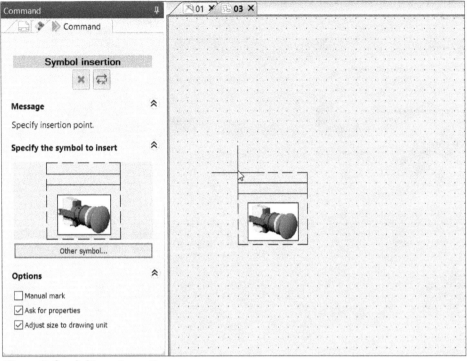

Figure-7. Symbol insertion options

- Click at desired location in the drawing to specify the insertion point. The **Symbol properties** dialog box will be displayed; refer to Figure-8.

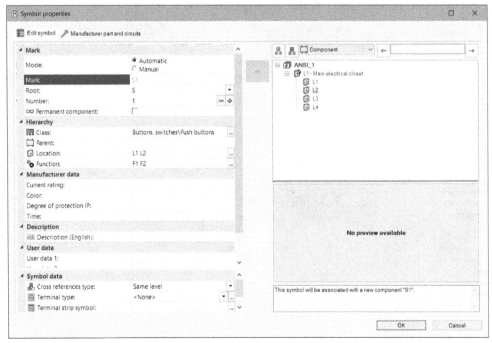

Figure-8. Symbol properties dialog box

- Specify desired manufacturer data in the fields of **Edit symbol** tab in the dialog box.

Manufacturer parts and Circuits

- To select component data from the SolidWorks Electrical library, click on the **Manufacturer part and circuits** tab. The dialog box will be displayed as shown in Figure-9.

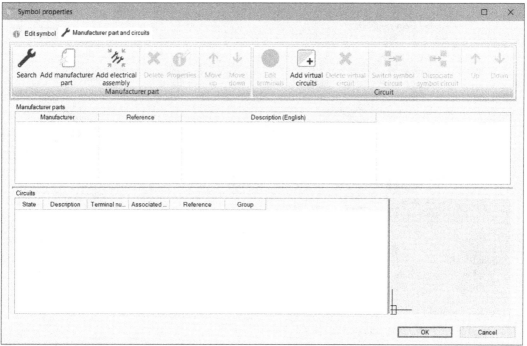

Figure-9. Symbol properties dialog box with manufacturer part and circuits

- Click on the **Search** button from the dialog box. The **Manufacturer part selection** dialog box will be displayed; refer to Figure-10.

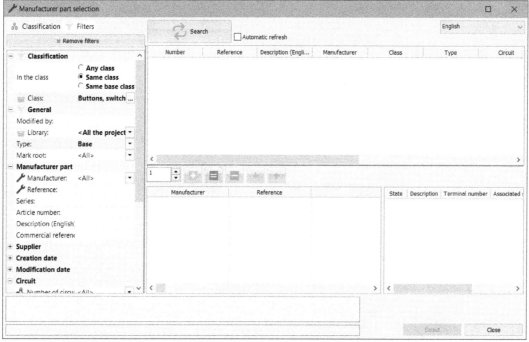

Figure-10. Manufacturer part selection dialog box

- Click on the ellipse button for **Class name** field under **Classification node** in the left of the dialog box; refer to Figure-11. The **Class selector** dialog box will be displayed; refer to Figure-12.

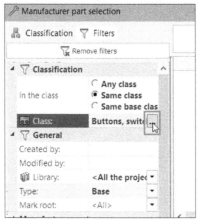

Figure-11. Ellipse button

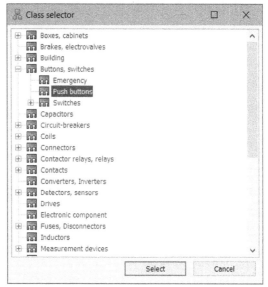

Figure-12. Class selector dialog box

- Select the class of component from the dialog box and click on the **Select** button.
- Click on the down arrow in the **Manufacturer** field of the **Manufacturer part** node and select the manufacturer of component; refer to Figure-13.

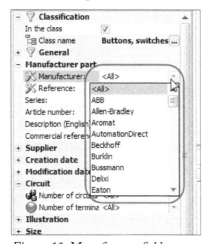

Figure-13. Manufacturer field

- Set the other filters as required and click on the **Search** button from the dialog box. The related manufacturer parts will be displayed; refer to Figure-14.
- Select desired component from the list and click on the **Add manufacturer part** button ⊞ from the dialog box. The selected component will be added in the current project; refer to Figure-15.

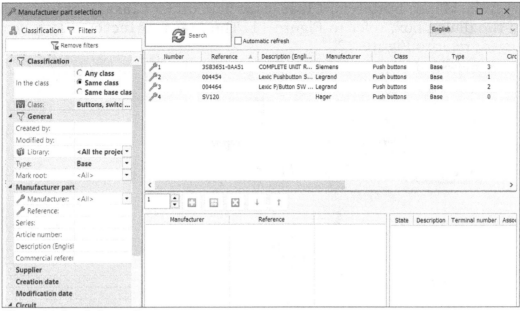

Figure-14. Searched manufacturer parts

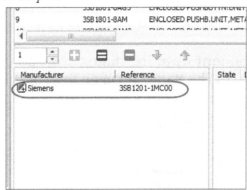

Figure-15. Part added in the project

- Similarly, you can add more manufacturer parts for the current project. Select the component from the new list of components added in the project and click on the **Select** button from the **Manufacturer part selection** dialog box. The **Symbol properties** dialog box will be displayed again.
- Select desired component manufacturing description from list displayed in the **Manufacturer part and circuits** tab of the dialog box and click on the **OK** button from the dialog box. The component will be placed with component description; refer to Figure-16.

Figure-16. Component placed

You can also insert symbols by using the **Symbols palette** displayed on the right in the application window; refer to Figure-17. If the **Symbols palette** is not displayed then click on the **Symbols palette** button from the **Dockable panels** panel in the **View** tab of the **Ribbon**; refer to Figure-18.

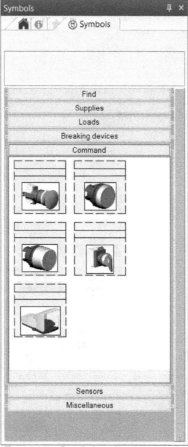

Figure-17. Symbols palette

Figure-18. Symbols Palette button

CONNECTION LABELS

A connection label is representation of a device in terms of connections. When we insert symbols by using the **Symbols** palette, we do not show the number of terminals of the part. To show the number of terminals along with the component symbol, we use the connection labels. A component which does not have a manufacturer part associated with it cannot be represented by a connection label. The procedure to insert a connection label is given next.

- Click on the **Insert a connection label for component** button from the **Connection label** drop-down in the **Insertion** panel of the **Ribbon**; refer to Figure-19. The **Command Browser** will be displayed with options related to labels and **Symbol selector** dialog box will be displayed; refer to Figure-20. Double-click on desired component in the dialog box. The component label will be attached to cursor; refer to Figure-21.

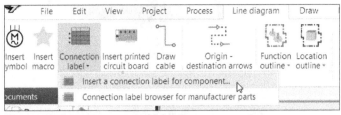

Figure-19. Insert a connection label for component button

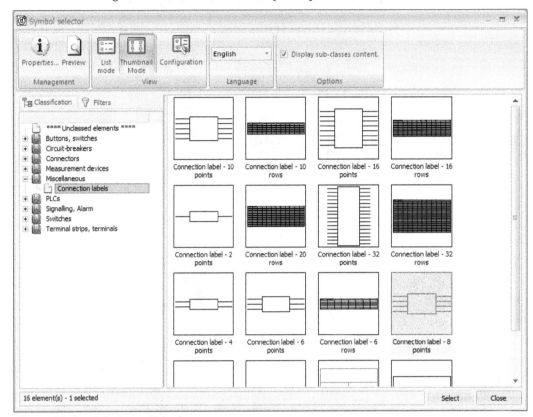

Figure-20. Symbol selector dialog box

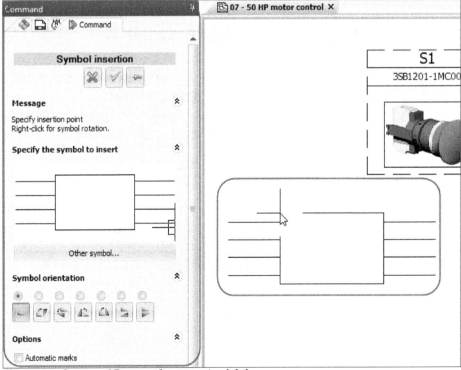

Figure-21. Command Browser for connection label

- Click on the **Other symbol** button from the **Command Browser** to select other symbol if you want to change the symbol.
- Select desired radio button from the **Symbol orientation** rollout in the panel to rotate the symbol; refer to Figure-22.

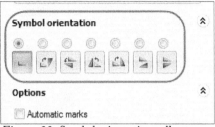

Figure-22. Symbol orientation rollout

- Click at desired location to place the symbol. The **Symbol properties** dialog box will be displayed as discussed earlier.
- Specify desired attributes to the symbol and click on the **OK** button.

Connection label browser for manufacturer parts

Earlier, we have learned to insert components in the project by using the Symbol palette. If you want to insert connection labels for the components earlier added in project then there is a direct method for that. The method is given next.

- Click on the **Connection label browser for manufacturer parts** button from the **Connection label** drop-down; refer to Figure-23. The **Connection label browser** will be displayed; refer to Figure-24.

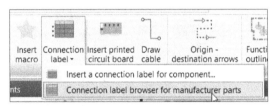

Figure-23. Connection label browser for parts button

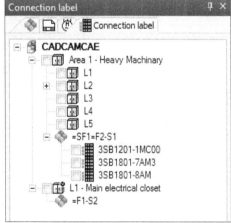

Figure-24. Connection label browser

- Click on the check box for desired component in the Connection label browser, the label for component will get attached to cursor.
- Click in the drawing to place it; refer to Figure-25.

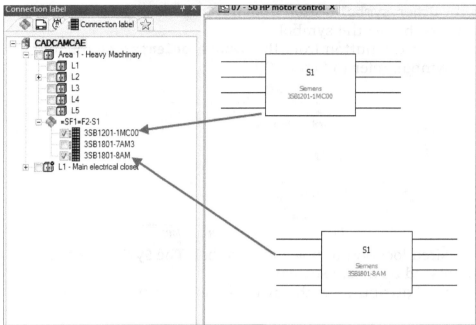

Figure-25. Connection labels placed

INSERTING REPORT TABLE

The **Insert Report Table** tool is used to insert bill of materials in the current drawing. The procedure to use this tool is given next.

- Click on the **Insert Report Table** tool from the **Insertion** panel in the **Line diagram** tab of the **Ribbon**. The **Report configuration selector** dialog box will be displayed; refer to Figure-26.

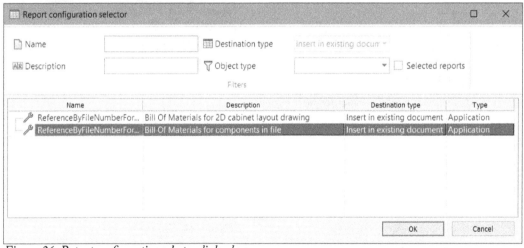

Figure-26. Report configuration selector dialog box

- Select desired report option from the list box and click on the **OK** button. The **Insert report table** page will be displayed in **Command Browser**.
- Click in the drawing area to specify start point and end point for defining width of the table. The table will be inserted in the drawing; refer to Figure-27.

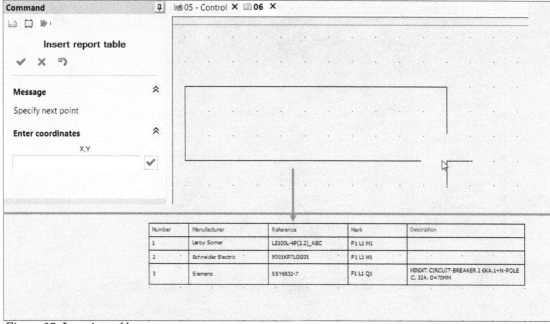

Figure-27. Inserting table

INSERTING PRINTED CIRCUIT BOARD

Sometimes in your electrical project, you may need to insert PCB (Printed Circuit Boards) to create desired functionality. In SolidWorks Electrical, you can insert PCB by using the **Insert printed circuit board** tool. The procedure to use this tool is given next.

- Click on the **Insert printed circuit board** tool from the **Insertion** panel in the **Line diagram** tab of the **Ribbon**. The **Printed circuit board insertion** dialog box will be displayed; refer to Figure-28.
- Click on the **Create a new printed circuit board from a manufacturer part** button if you have printed circuit boards in SolidWorks Electrical library and want to use them. Click on the **Create a new printed circuit board from an electronic design file** button if you want to use a electronic design file created by electronic design software.

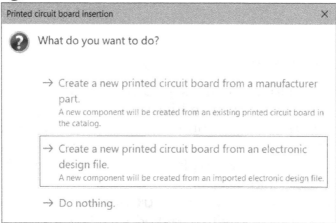

Figure-28. Printed circuit board insertion dialog box

- In our case, we have selected **Create a new printed circuit board from an electronic design file** button. On doing so, the **Select printed circuit board file** dialog box will be displayed; refer to Figure-29.

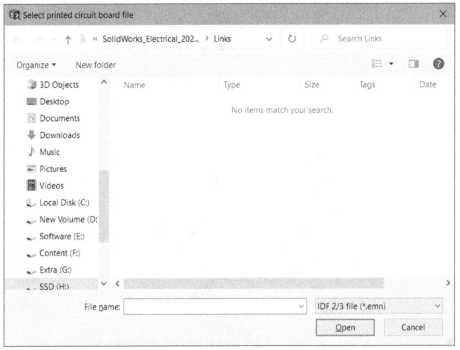

Figure-29. Select printed circuit board file dialog box

- Select the circuit file with `.emn` extension and click on the **Open** button from the dialog box. A SolidWorks message box will be displayed asking you whether to copy the file or let the system use original file; refer to Figure-30.

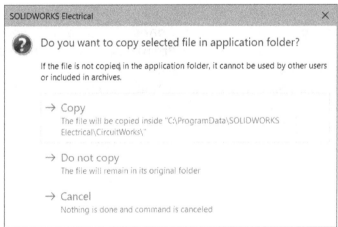

Figure-30. SolidWorks Electrical message box

- Click on desired button from the message box. The **Manufacturer Part properties** dialog box will be displayed; refer to Figure-31.
- Set desired parameters and click on the **OK** button from the dialog box (You will learn about the options in this dialog box later in this book). The **Component properties** dialog box will be displayed (You will learn about options in this dialog box later in the book).
- Set desired parameters and click on the **OK** button. The **Symbol selector** dialog box will be displayed and you will be asked to select a symbol for circuit board.
- Double-click on desired symbol in the dialog box. The symbol will get attached to cursor.
- Click at desired location to place the symbol.

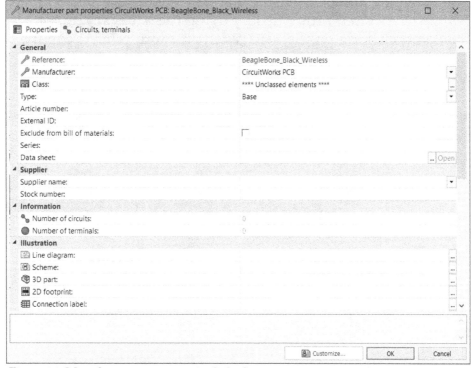

Figure-31. Manufacturer part properties dialog box

DRAWING CABLES

Once you have inserted desired components, the next step is to connect them with the help of a cable. The procedure to draw cable is given next.

• Click on the **Draw cable** button from the **Insertion** panel in the **Line diagram** tab of the **Ribbon**. The **Command Browser** will be displayed with the options related to cable; refer to Figure-32.

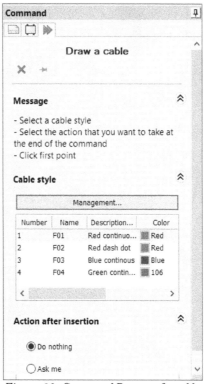

Figure-32. Command Browser for cable

- Select desired wire number from the list in the **Command Browser**. You can also create a wire with desired properties by using the **Wire style management** dialog box displayed on clicking on the **Management** button; refer to Figure-33.

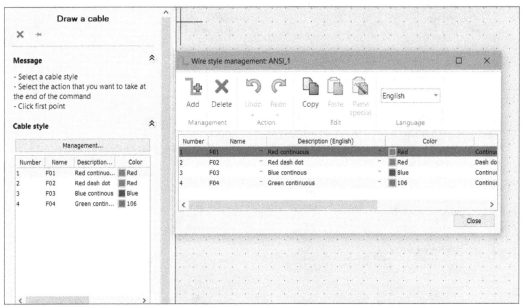

Figure-33. Wire style management dialog box

- Click on the **Add** button from the **Wire style management** dialog box and add the wire style with desired color and description.
- Click on the **Close** button from the **Wire style management** dialog box to close the dialog box and click on the component boundary to specify starting point of the wire; refer to Figure-34.

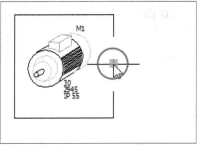

Figure-34. Starting point of wire

- Move the cursor and click at desired location to make bend in the wire; refer to Figure-35.

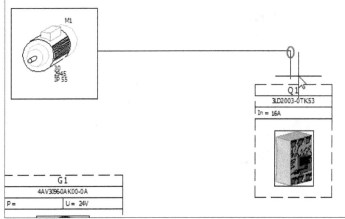

Figure-35. Point clicked to make bend

- Move the cursor upward/downward and click to specify the end point of the wire connected to the component. The **Cable insertion** dialog box will be displayed if **Ask me** radio button is selected in the **Action after insertion** node of **Draw a cable Command Manager**; refer to Figure-36.

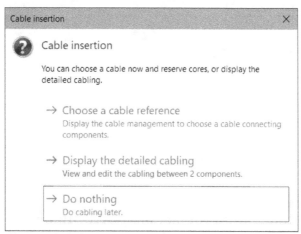

Figure-36. Cable insertion dialog box

- We will discuss about the **Display the detailed cabling** option later in this chapter. Click on the **Choose a cable reference** option from the dialog box. The **Cable references selection** dialog box will be displayed; refer to Figure-37. You can also display the same dialog box by clicking on the **Add a cable** tool from the **Changes** panel in the **Line diagram** tab of **Ribbon**.

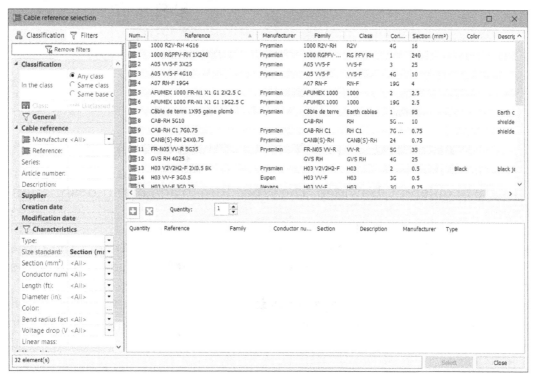

Figure-37. Cable references selection dialog box

- Select desired cable from the list displayed at the top of the dialog box and click on the **Add** button; refer to Figure-38.

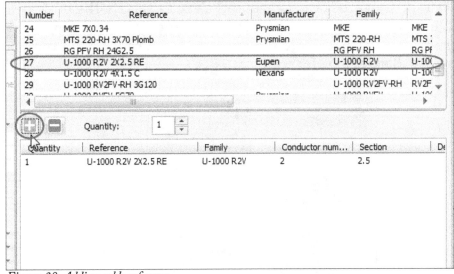

Figure-38. Adding cable reference

- Select the cable reference from the list of added cable references and click on the **Select** button from the dialog box. The reference will be attached to the cable drawn; refer to Figure-39.

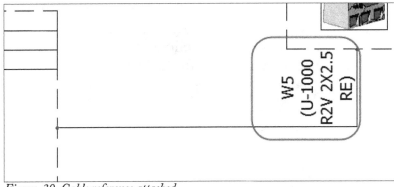

Figure-39. Cable reference attached

ORIGIN - DESTINATION ARROWS

The origin-destination arrows are used to link the components sharing same cable or wire but are in different drawings. The procedure to add origin-destination arrows is given next.

- Click on the **Origin - destination arrows** tool from the **Insertion** panel in the **Line diagram** tab of the **Ribbon**. The **Origin-destination manager** will be displayed; refer to Figure-40.
- Using the options in the **Change Drawing** panels, you can switch between the drawings being displayed if you want to use different from being displayed by default; refer to Figure-41.

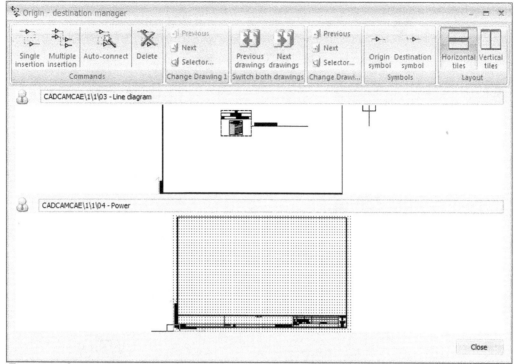

Figure-40. Origin-destination manager

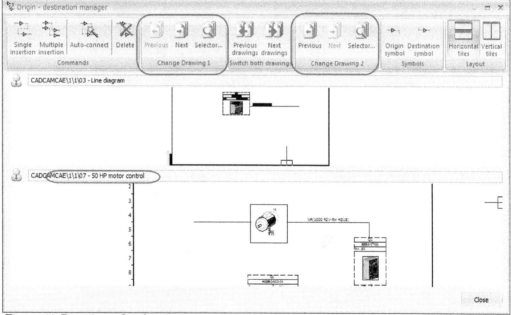

Figure-41. Drawings after changing

- Click on the **Auto-connect** button from the **Commands** panel if the cables are of same reference numbers. The cables with same numbers and properties will get connected automatically.
- To manually set the link, click on the **Single insertion** button from the **Management** panel in the **Ribbon** of dialog box. Open end will automatically get highlighted in the drawing; refer to Figure-42.
- Click on the end highlighted by green circle in the first drawing. It will turn into red circle and you will be asked to selected the other end point.
- Select the end point of cable/wire in other drawing; refer to Figure-43. Arrow heads will get attached to the cursor; refer to Figure-44.

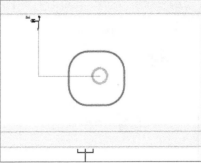

Figure-42. Highlighted open end

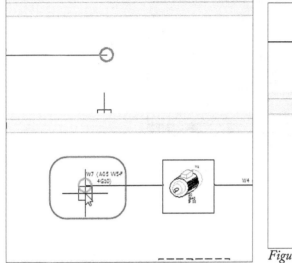

Figure-43. Selecting other end point

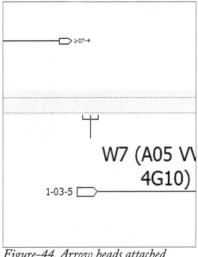

Figure-44. Arrow heads attached

- To change the arrow head style, click on the **Origin Symbol** or **Destination Symbol** button (as required) from the **Symbols** panel in the **Origin-destination manager**. The respective dialog box will be displayed; refer to Figure-45.

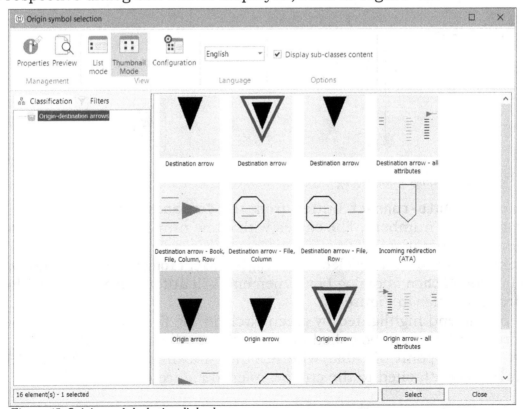

Figure-45. Origin symbol selection dialog box

- Select desired style from the templates and click on the **Select** button.
- In the same way, you can connect other cables/wires in different drawings.

FUNCTION OUTLINE

The Function outlines are used to mark the components on the basis of their functions in the circuit. These outlines are just to categorize the components in drawing on the basis of function. The procedure to create function outlines is given next.

- Click on the down arrow below **Function outline** button in the **Insertion** panel of the **Line diagram** tab in the **Ribbon**. The tools for outlining will be displayed; refer to Figure-46.

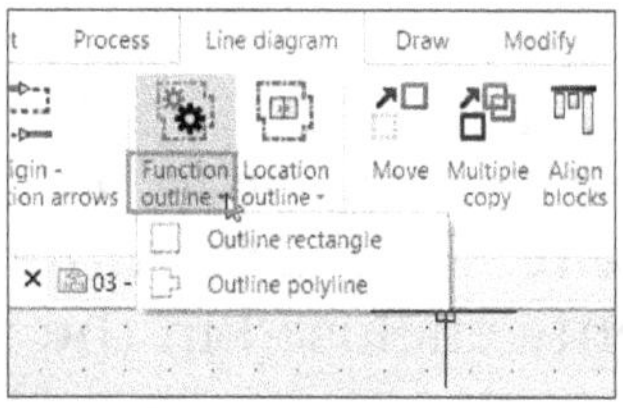

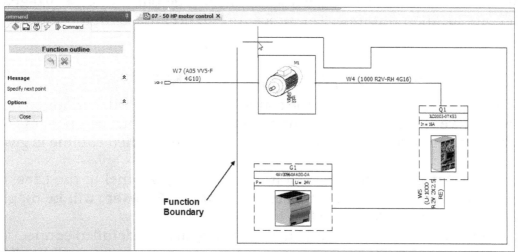

Figure-46. Tools for function outlining

- Select desired tool from the list displayed (**Outline polyline** selected in our case) and drawing a boundary for the components of same function; refer to Figure-47.

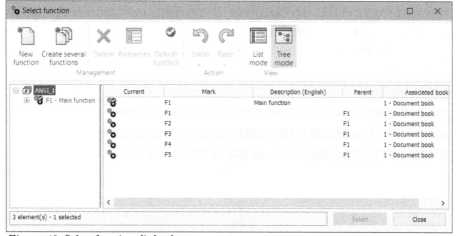

Figure-47. Function outlining created

- Click on the **Close** button from the **Command Browser** to close the polyline boundary. The **Select function** dialog box will be displayed; refer to Figure-48.

Figure-48. Select function dialog box

- Select desired function from the list and click on the **Select** button from the dialog box. The **Change component function** dialog box will be displayed; refer to Figure-49.

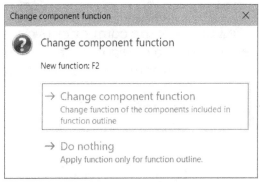

Figure-49. Change component function dialog box

- Click on the **Change component function** button from the dialog box to change the function of the components enclosed in the boundary.

LOCATION OUTLINE

The Location outlines are used to mark the component on the basis of their locations in the circuit. The **Location outline** tool in **Insertion** panel of **Line diagram** tab is used to create location outlines. The procedure to create location outline is similar to creating the Function outline.

DETAILED CABLING

Detailed cabling is used to represent the connections of cable with various components. In other words, we can describe the connection between various components of circuit with the help of detailed cabling. The procedure to use detailed cabling is given next.

- Click on the **Detailed Cabling** tool from the **Changes** panel in the **Line diagram** tab of the **Ribbon**. The **Detailed cabling Command Browser** will be displayed at the left and you are asked to select a cable.
- Click on desired cable for which you want to specify the detailed connections and press **ENTER** from keyboard. The **Detailed cabling** dialog box will be displayed; refer to Figure-50.

The dialog box is divided into three sections; Origin component, Cable, and Destination component; refer to Figure-51. To connect the origin component, use the options in **Origin component** section and similarly for the **Destination component**. Note that the numbers under screw columns in origin and destination component sections display the terminal numbers. Procedure to perform connection by cable is given next.

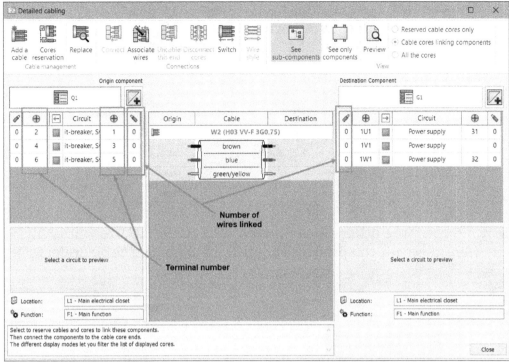

Figure-50. Detailed cabling

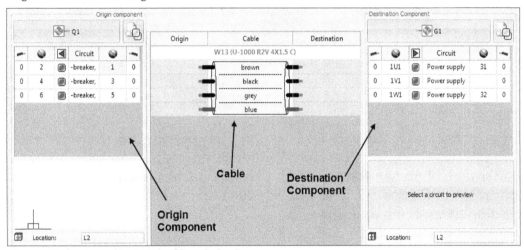

Figure-51. Sections of Detailed cabling dialog box

- Click in the wire box next to terminal 1 in the **Origin component** section and then click on the left green box for brown wire; refer to Figure-52.

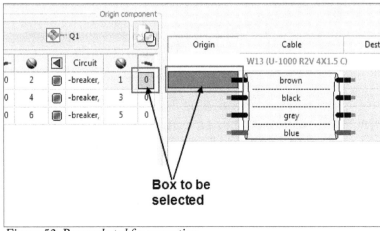

Figure-52. Boxes selected for connection

- Click on the **Connect** button from the **Connections** panel in the dialog box. The connection will be created as shown in Figure-53. Here, **Q1** is name of origin component and **1** is the terminal number in **Q1:1**.

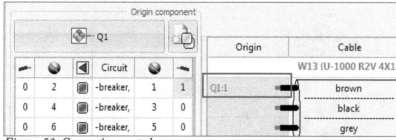

Figure-53. Connection created

- Similarly, connect the other side of brown wire to the 1U1 terminal of destination component and repeat the process for other wires in the cable; refer to Figure-54.

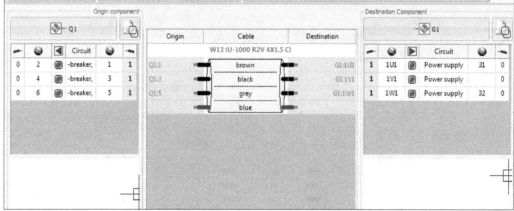

Figure-54. Connection with destination component

- Now, we are left with blue wire and we don't have terminals to connect it with. We are going to use this wire as ground. To make a terminal for ground in the Origin component, click on the **Add virtual circuit** button from the **Origin component** section; refer to Figure-55. The **Add virtual circuits to component** dialog box will be displayed; refer to Figure-56.

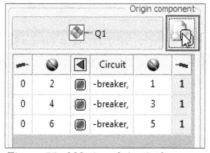

Figure-55. Add virtual circuits button

- Click on the **Add** button from the dialog box. A new circuit will be added in the list. Click on the **More circuit types** option from the drop-down list displayed on clicking down arrow in **Circuit type** column; refer to Figure-57.

Figure-56. Add virtual circuits to component dialog box

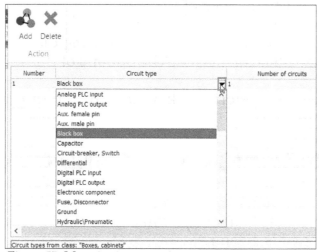

Figure-57. More circuit types option

- Click on the **Ground** option from the drop-down list; refer to Figure-58. Specify the **Number of circuits** as **1** and **Number of terminals per circuit** as **2**.

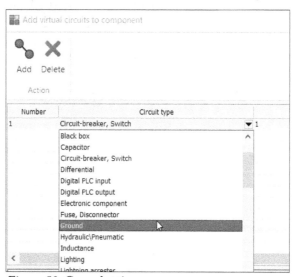

Figure-58. Ground option

- Click on the **OK** button from the dialog box.
- Connect the ground with the blue wire and do the same procedure on other side; refer to Figure-59.

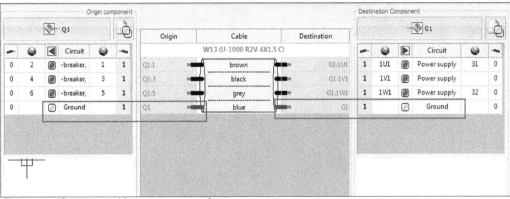

Figure-59. Connecting blue wire to grounds

- To disconnect cable or cores, click on the **Uncable this end** or **Disconnect cores** button from the dialog box, respectively. Selecting the **Uncable this end** button will remove connection of selected side of cable. Selecting the **Disconnect cores** button will disconnect both sides of the wire.
- After selecting one core of wire, you can change its wire style by clicking on the **Wire style** button from the **Detailed cabling** dialog box. The **Wire style selector** dialog box will be displayed. The options of this dialog box have already been discussed.
- After setting desired parameters, click on the **Close** button from top-right corner to exit the dialog box.

RESERVING CABLES AND CORES

The **Reserve cables and cores** tool is used to reserve selected cable or its cores for a specific component. By doing so, you make sure that these cables/cores are not used accidentally for other component while performing wiring connections. The procedure to use this tool is given next.

- Click on the **Reserve cables and cores** tool from the **Changes** panel of **Line diagram** tab in the **Ribbon**. The **Cables reservation Command Manager** will be displayed.
- Select the cable whose cores are to be reserved and click on the **OK** button from the **Command Manager**. The **Cables and cores reservation** dialog box will be displayed; refer to Figure-60.

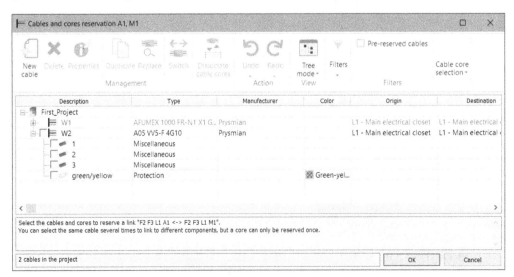

Figure-60. Cables and cores reservation dialog box

- Select desired check boxes of cores for cable which are to be reserved; refer to Figure-61.

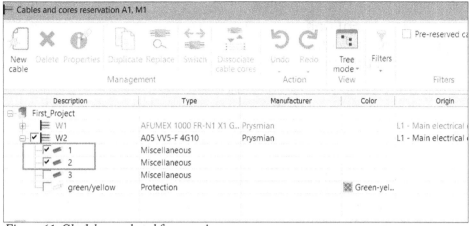

Figure-61. Check boxes selected for reserving cores

- After selecting desired cores/cables, click on the **OK** button from the dialog box.

MOVING COMPONENT

The **Move** tool in **Changes** panel of **Ribbon** is used to move a component from one location to another without removing connections. The procedure to use this tool is given next.

- Click on the **Move** tool from the **Changes** panel of **Line diagram** tab in the **Ribbon**. The **Move Command Manager** will be displayed and you will be asked to select the component to be moved.
- Select desired component from the drawing area and press **ENTER**. You will be asked to specify base point for moving component as reference.
- Click at desired location on the component to specify the base point. You will be asked to specify target location where you want to move the selected component.
- Click at desired location in the drawing to place the component; refer to Figure-62.

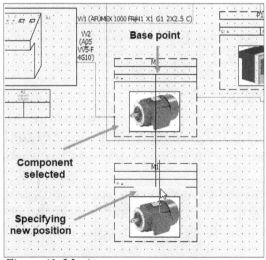

Figure-62. Moving component

CREATING MULTIPLE COPIES OF OBJECTS

The **Multiple Copy** tool is used to create multiple copies of selected component/ object in the drawing. The procedure to use this tool is given next.

- Click on the **Multiple Copy** tool from the **Changes** panel in the **Line diagram** tab of the **Ribbon**. The **Copy Command Manager** will be displayed and you will be asked to select the component to be copied.
- Select desired component and press **ENTER**. You will be asked to specify base point for placing copies of the component.
- Click at desired location to specify base point. A copy of the component will get attached to cursor.
- Click at desired location to place the copy. You will be asked to specify placement location for next copy.
- Place desired number of copies and then press **ESC** to exit the tool.

ALIGNING BLOCKS

The **Align Blocks** tool is used to align selected components along the insertion point of selected reference object. The procedure to use this tool is given next.

- Click on the **Align Blocks** tool from the **Changes** panel in the **Line diagram** tab of the **Ribbon**. You will be asked to select the component to be used as reference for alignment.
- Select desired reference object/component and then press **ENTER**. The **Align blocks Command Manager** will be displayed as shown in Figure-63.

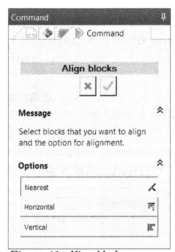

Figure-63. Align blocks CommandManager

- Select desired button from the **Options** rollout in **Manager** to specify which orientation will be used for alignment. Select the **Horizontal** button to horizontally align selected components and select the **Vertical** button to vertically align selected components.
- After selecting desired button, select multiple components near their insertion points while holding the **CTRL** key and then click on the **OK** button from the **Manager**. The selected components will be aligned according to parameters specified.

Chapter 4

Schematic Drawing

Topics Covered

The major topics covered in this chapter are:

- *Introduction*
- *Starting Schematic Drawings*
- *Inserting Symbols, Wires, and Black boxes*
- *Inserting PLCs, Connectors, and Terminals*
- *Inserting Reports*
- *Performing Design Rule Checks*

INTRODUCTION

As discussed earlier in the book, schematic drawing is the drawing showing all significant components and parts of a circuit with their interconnections. The tools to create schematic drawing are available in the **Schematic** tab of **Ribbon**; refer to Figure-1. Note that this tab will be available only when you are working on a schematic drawing. The procedure to start a schematic drawing in project is given next.

Figure-1. Schematic tab

STARTING A SCHEMATIC DRAWING

* Click on the **New scheme** tool from the **New** drop-down in the **Electrical Project** panel of the **Electrical Project** tab in the **Ribbon**; refer to Figure-2. A new drawing will be added to the document book.
* Right-click on the newly added drawing in the **Pages Browser** and select the **Properties** option from the shortcut menu. The **Drawing** dialog box will be displayed; refer to Figure-3.
* Click in the **Description** field and specify the description for drawing.
* Set the location & function of the drawing and click on the **OK** button from the dialog box to start the drawing. The drawing will be modified accordingly.
* Double-click on the drawing file name in the **Pages Browser** to open it if not opened earlier. The tools related to schematic drawing will be displayed; refer to Figure-4.

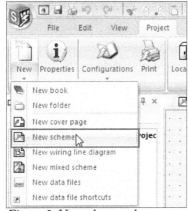

Figure-2. New scheme tool

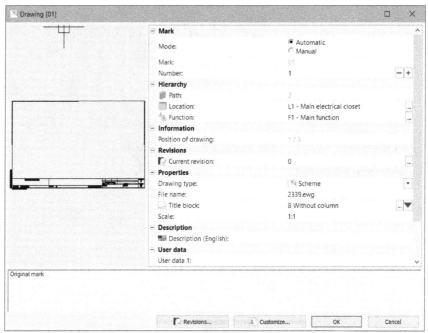

Figure-3. Drawing dialog box

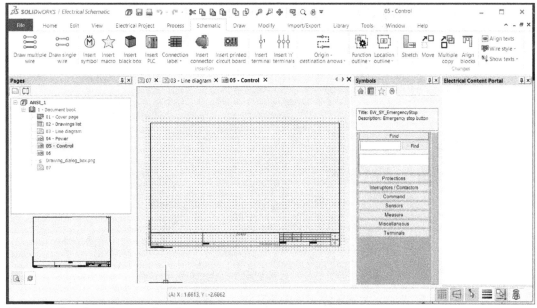

Figure-4. Starting schematic drawing

INSERTING SYMBOL

The procedure to insert symbols in schematic drawing is similar to the procedure of inserting symbol in line diagrams. The procedure to insert schematic symbols is given next.

*   Click on the **Insert symbol** tool from the **Insertion** panel in the **Schematic** tab of the **Ribbon**. If the **Symbol insertion** page is displayed in the **Command Browser** as shown in Figure-5 then click on the **Other symbol** button from the **Command Browser**. The **Symbol selector** dialog box will be displayed; refer to Figure-6. Otherwise, the **Symbol selector** dialog box will be displayed directly.

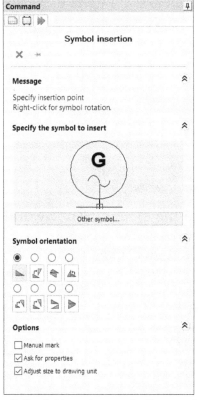

Figure-5. Symbol insertion page in Command Browser

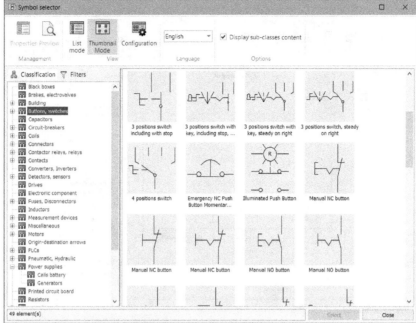

Figure-6. Symbol selector dialog box

- Select desired symbol from the dialog box and click on the **Select** button. The symbol will get attached to cursor.
- Select desired orientation for the symbol from the **Symbol orientation** area in the **Command Browser**; refer to Figure-7.

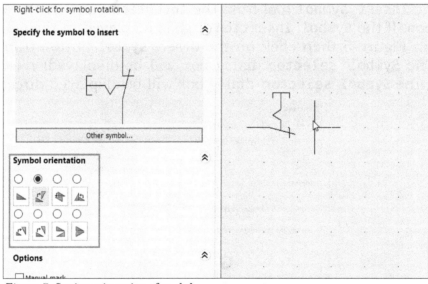

Figure-7. Setting orientation of symbol

- Click in the drawing to place the symbol. The **Symbol properties** dialog box will be displayed; refer to Figure-8.
- Click on the **Manufacturer part and circuits** tab and set the manufacturer data for component. The options in the dialog box have been discussed earlier in the book.
- Click on the **OK** button from the dialog box. The symbol will be displayed along with its attributes; refer to Figure-9.

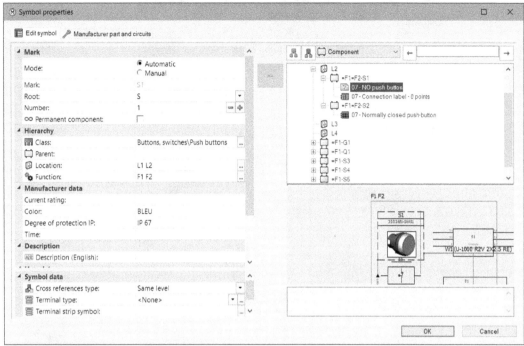

Figure-8. Symbol properties dialog box

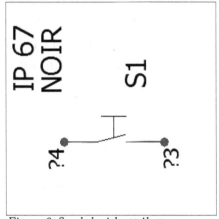

Figure-9. Symbol with attributes

INSERTING WIRES

We have wires for schematic drawings like, cables for Line diagrams. We have two tools named **Draw multiple wire** and **Draw single wire** to insert wires in the schematic drawings. The procedures to create both the wires are given next.

Inserting Single Wire

Wires are used to connect the components so that they can function as required. The procedure to insert single wire in the drawing is discussed next.

- Click on the **Draw single wire** tool from the **Insertion** panel of the **Schematic** tab in the **Ribbon**. The **Electrical wires** page will be displayed in the **Command Browser**; refer to Figure-10.

Figure-10. Electrical wires page in Command Browser

- Click on the ellipse button [...] next to **Name** field in the **Wire style selection** area of the **Command Browser**. The **Wire style selector** dialog box will be displayed; refer to Figure-11.

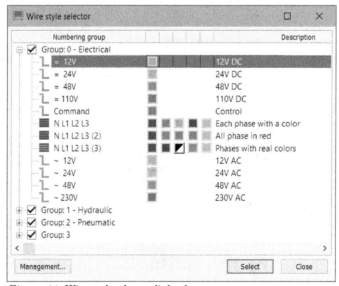

Figure-11. Wire style selector dialog box

- Select desired wire from the list in the dialog box.
- If desired wire is not available in the list, click on the **Management** button from the dialog box. The **Wire style management** dialog box will be displayed; refer to Figure-12.

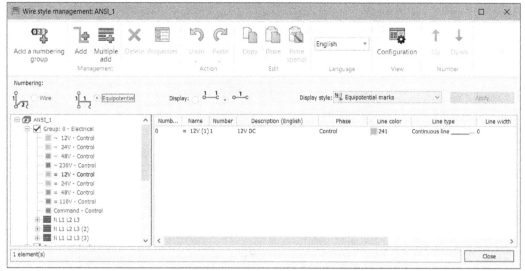

Figure-12. Wire style management dialog box

- Create a wire style with desired parameters as discussed in previous chapter. Select the newly created wire style.
- Click on the **Select** button from the **Wire style selector** dialog box. You are asked to specify the start point of the wire.
- Click to specify the starting point. You are asked to specify the next point of the wire. Press **F8** if you want to create non-ortho wire.
- Click to specify the corners of wire and press **ENTER** when you want to exit the wire creation but still want wire creation mode active. Press **ESC** to exit the tool.
- If you want to create more than one wires then set desired number in the **Number of lines** spinner of the **Electrical wires** page in **Command Browser**; refer to Figure-13. Specify desired value in **Space between lines** edit box to define space between two consecutive wires. Rest of the procedure is same as discussed.

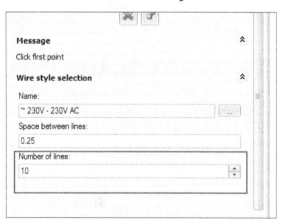

Figure-13. Number of lines spinner

Inserting Multiple Wires

For electrical supplies like three phase connection, we need a set of four wires. Such connections can be made by using the **Draw multiple wire** tool. The procedure to use this tool is given next.

- Click on the **Draw multiple wire** tool from the **Insertion** panel of the **Ribbon**. The **Electrical wires** page will be displayed in the **Command Browser**. If you are using this tool for the first time then the **Wire style selector** dialog box will be displayed. Select desired wire from the dialog box. Note that the wire with the multiple wire icon are used to draw multiple wires.

• Select desired check box from the **Available wires** area to enable wires in the wire set; refer to Figure-14.

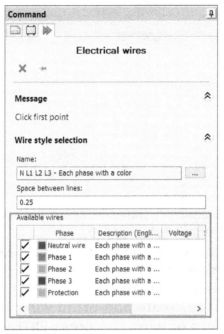

Figure-14. Available wires area

• You can specify the distance between two consecutive wire lines by using the **Space between lines** edit box.
• After setting desired parameters, click in the drawing area to specify the starting point for the wires. The procedure of drawing multiple wires is same as for single wire.

INSERTING BLACK BOX

Black box is an undefined object in SolidWorks Electrical. We can use black box when we don't want to use library symbol or we don't have the symbol for our component. Black box has a specific property that when you place a black box on any wire or set of wires, the respective number of terminals are automatically created on it. The procedure to insert black box in drawing is given next.

• Click on the **Insert black box** tool from the **Insertion** panel in the **Schematic** tab of the **Ribbon**. If you have selected a symbol of black box earlier then the **Symbol insertion** page will be displayed in the **Command Browser** as shown in Figure-15.
• If you are using this tool for the first time then the **Symbol selector** dialog box will be displayed; refer to Figure-16.
• Select desired symbol and click on the **Select** button from the dialog box.
• Click in the drawing to specify the starting point of black box. You are asked to specify the other corner point for rectangular boundary of black box; refer to Figure-17.

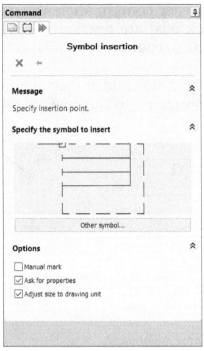

Figure-15. Symbol insertion page

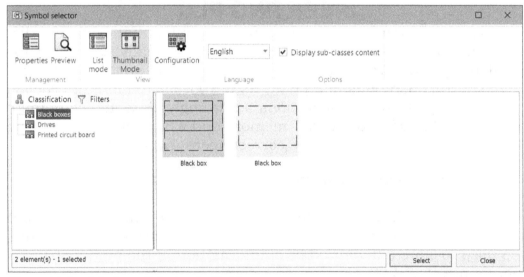

Figure-16. Symbol selector dialog box for black box

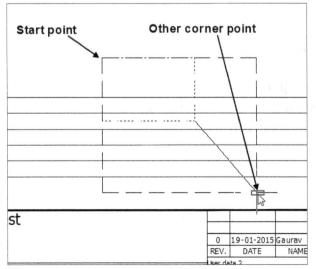

Figure-17. Creating boundary of black box

- Click to specify the other corner point in such a way that desired wires (which you want to connect with black box) are overlapped by the rectangle. On specifying the corner point, the **Symbol properties** dialog box will be displayed as discussed earlier.
- Set desired properties in the dialog box and click on the **OK** button from the dialog box. The symbol will be connected to the covered wires; refer to Figure-18.

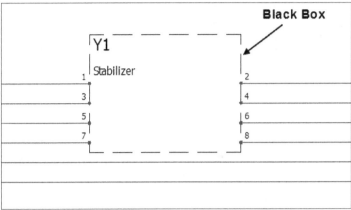

Figure-18. Black box symbol placed

INSERTING PLC

PLC is a solid state/computerized industrial computer that performs discrete or sequential logic in a factory environment. It was originally developed to replace mechanical relays, timers, counters. PLCs are used successfully to execute complicated control operations in a plant. Its purpose is to monitor crucial process parameters and adjust process operations accordingly. A sequence of instructions is programmed by the user to the PLC memory and when the program is executed, the controller operates a system to the correct operating specifications.

PLC consists of three main parts: CPU, memory and I/O units.

CPU is the brain of PLC. It reads the input values from inputs, runs the program existing in the program memory and writes the output values to the output register. Memory is used to store different types of information in the binary structure form. The memory range of S7-200 is composed of three main parts as program, parameter, and retentive data fields. I/O units provide communication between PLC control systems.

The procedure to insert PLC in drawing is discussed next.

- Click on the **Insert PLC** tool from the **Insertion** panel of the **Schematic** tab in the **Ribbon**. The **Dynamic PLC insertion** dialog box will be displayed if there is another PLC already inserted in the drawing; refer to Figure-19.

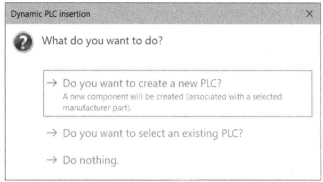

Figure-19. Dynamic PLC insertion dialog box

Creating New PLC

- Click on the **Do you want to create a new PLC?** button from the dialog box to create a new PLC. The **Manufacturer part selection** dialog box will be displayed; refer to Figure-20.

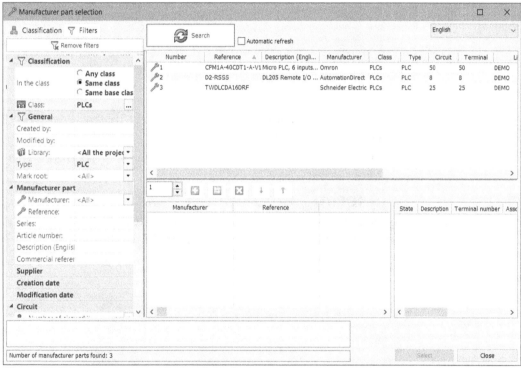

Figure-20. Manufacturer part selection dialog box

- Search the PLC with desired number of terminals and properties. Click on the **+** button to add it in project.
- Click on the **Select** button to select it. The **Component properties** dialog box will be displayed.
- Specify the user data as discussed earlier in the dialog box and click on the **OK** button. PLC will get attached to cursor and parameters related to channels will be displayed in the **Command Browser**; refer to Figure-21.

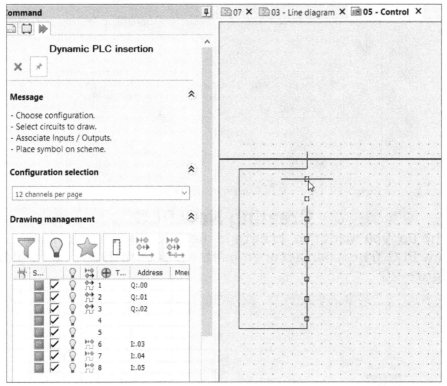

Figure-21. Parameters related to channels

- Select the check boxes from the **Channel selection** area in the **Command Browser** to enable respective connection points.
- Similarly, set the other parameters and click in the drawing to place the PLC.

Inserting an existing PLC

- If there is an existing PLC in the project but not created yet in schematic then select the **Do you want to select an existing PLC?** button from the **Dynamic PLC insertion** dialog box. The **PLC Selector** dialog box will be displayed; refer to Figure-22.

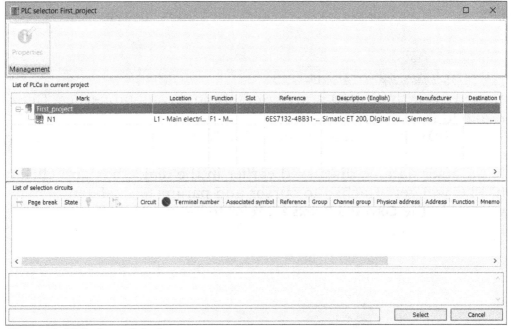

Figure-22. PLC selector dialog box

- Select desired PLC from the list and click on the **Select** button. The PLC will get attached to cursor.
- Click at desired location to place the PLC.

You can insert Connection label, Report table, and Printed Circuit Board as discussed in previous chapter.

INSERTING CONNECTORS

Connectors are used to facilitate easy assembly of components to the circuit. The most common example of connectors can be the power outlet and power plug. There is a big list of connectors available in the market so we are not going to discuss details of them in this book. The procedure to insert all the types of connectors is same and is given next.

- Click on the **Insert Connector** tool from the **Insertion** panel in the **Schematic** tab of the **Ribbon**. The **Manufacturer part selection** dialog box will be displayed.
- Search the connector with desired number of terminals by using filters and add it to the project.
- Click on the **Select** button from the dialog box. The **Component properties** dialog box will be displayed as discussed earlier.
- Set desired parameters and click on the **OK** button from the dialog box. The **Connector dynamic insertion** page of **Command Browser** will be displayed and connector will get attached to cursor; refer to Figure-23.

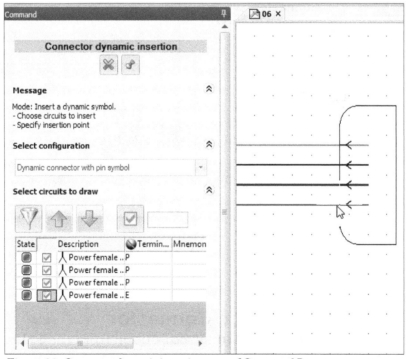

Figure-23. Connector dynamic insertion page of Command Browser

- Select desired option from the **Select configuration** drop-down in the panel. If you select the **Dynamic connector with pin symbol** option then connector will be displayed with pins; refer to Figure-24. If you select the **Dynamic connector without pin symbol** option then connector will be displayed without pins; refer to Figure-24. If you have selected the **One symbol per pin** option then connector symbol will not be displayed but pin symbol will be applied to every selected connection; refer to Figure-24. If you want to insert a symbol available in the **Symbol selector** dialog box then select the **<Select symbol to insert>** option from the **Select configuration** drop-down and click on the **Other symbol** button from the panel. The **Symbol Selector** dialog box will be displayed.

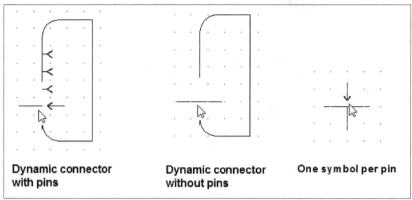

Figure-24. Connector options

- Select desired symbol and click on the **OK** button from the dialog box. The symbol will get attached to the cursor; refer to Figure-25.
- Hover the cursor to a terminal till you get a snap point and then click, the connector will automatically create connections with the terminal. Refer to Figure-26 in which wires are connected to the connector. Note that if all the connections are not aligned with connector then you can select the **Force associations** option from the **Circuit association** dialog box.

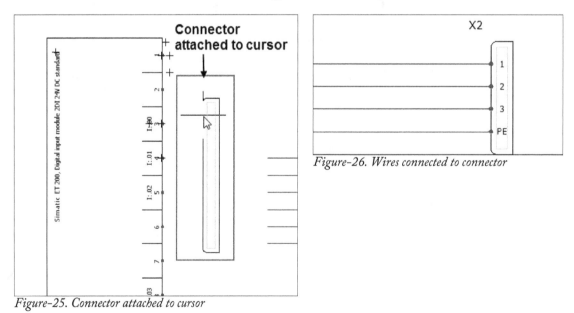

Figure-26. Wires connected to connector

Figure-25. Connector attached to cursor

Dynamic Connector

- Select the **Dynamic connector with pin symbol** option from the **Select configuration** drop-down. You will be asked to specify the insertion point of the connector.
- Click on the wire/terminal that you want to connect. The connector will be created; refer to Figure-27.

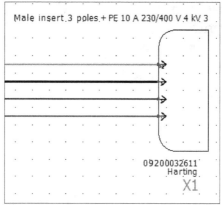

Figure-27. Connector created by dynamic connector option

You will learn about customizing connector in the next chapter.

INSERTING TERMINAL /TERMINALS

Terminals are used to allow connection of wires to the main circuit. In other words, terminals allow branch circuits to be connected with the main circuit. The procedure to insert terminal/terminals is given next.

Inserting single terminal

*   Click on the **Insert terminal** button from the **Insertion** panel in the **Schematic** tab of the **Ribbon**. The **Terminal selector** dialog box will be displayed (for the first time users); refer to Figure-28.

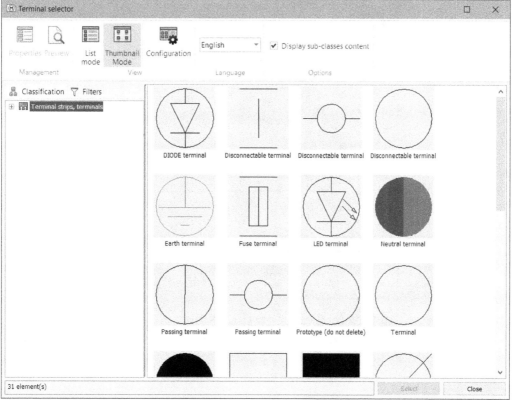

Figure-28. Terminal selector dialog box

*   Select desired terminal symbol and click on the **Select** button from the dialog box. The symbol will get attached to the cursor.

- Click on the wire to insert terminal; refer to Figure-29. You are asked to specify the orientation of the terminal.

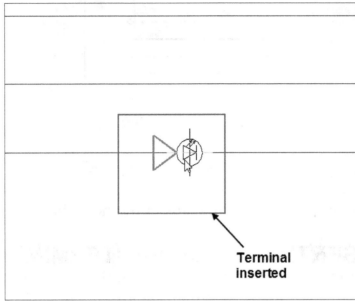

Figure-29. Terminal inserted

- Click on desired side of terminal to specify the orientation of the terminal. The **Terminal symbol properties** dialog box will be displayed; refer to Figure-30.

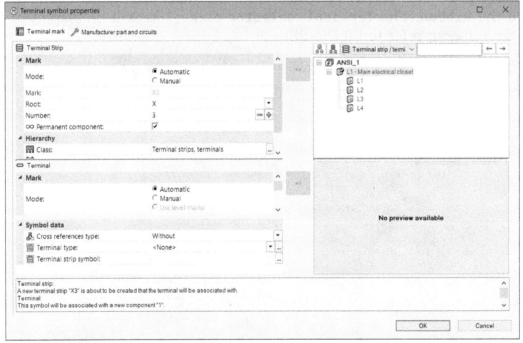

Figure-30. Terminal symbol properties dialog box

- Specify desired parameters and click on the **OK** button from the dialog box to create the terminal. If you want to associate a new terminal to terminals earlier created then follow the steps given in Figure-31 and click on the **OK** button.

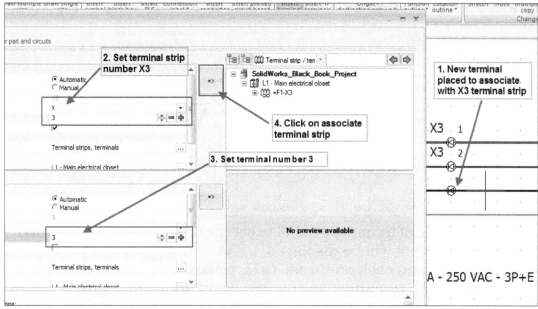

Figure-31. Associating new terminal to strip

Inserting multiple terminals

- Click on the **Insert 'n' terminals** button from the **Insertion** panel in **Schematic** tab of the **Ribbon**. If you are using the tool for the first time then **Terminal selector** dialog box will be displayed. Double-click on desired symbol from the **Terminal selector** dialog box. The **Terminal insertion** page will be displayed in the **Command Browser**; refer to Figure-32 and you will be asked to draw an axis line intersecting with wires for creating terminals.

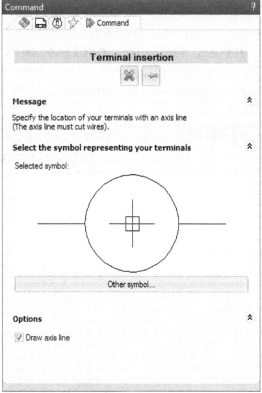

Figure-32. Terminal insertion page

- Click on the **Other symbol** button and select the symbol as discussed earlier if you want to change the terminal symbol.

• Draw an axis intersecting the wires to create terminals; refer to Figure-33.

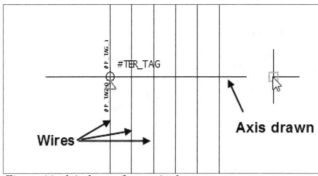

Figure-33. Axis drawn for terminals

• Click at desired side of terminal symbol to set the orientation. The **Terminal symbol properties** dialog box will be displayed.

• Set the properties and click on the **OK (all terminals)** button to apply the same properties to all the terminals or you can apply individual properties by using the **OK** button. The terminals will be created in the form of a strip aligned to axis; refer to Figure-34.

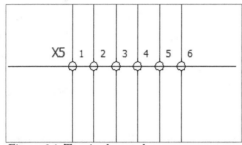

Figure-34. Terminal created

Till this point, we have covered almost all the tools related to schematics. The tools like **Origin-destination arrows**, **Function outline**, and **Location outline** have already been discussed in previous chapters. Now, we will discuss about editing terminal strips and inserting reports in the project.

TERMINAL STRIP EDITOR

Terminals are the connecting points used for various circuits so it is important to manage terminals properly. There is a special tool to manage connections of terminals at one place. After creating terminal, follow the procedure given next to manage terminal strip.

• Click on the **Terminal strips** tool from the **Management** panel in the **Electrical Project** tab of **Ribbon**. The **Terminal strips manager** dialog box will be displayed; refer to Figure-35.

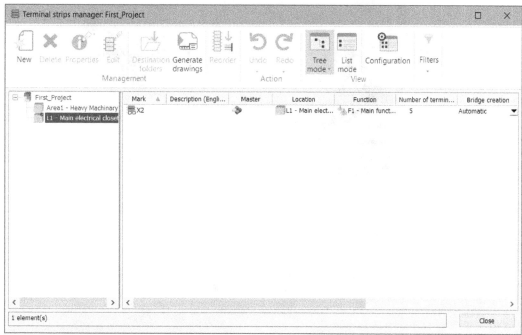

Figure-35. Terminal strips manager dialog box

- List of terminals used in the current project will be displayed in the dialog box. Double-click on the terminal strip from the dialog box. The **Terminal strip editor** dialog box will be displayed; refer to Figure-36.

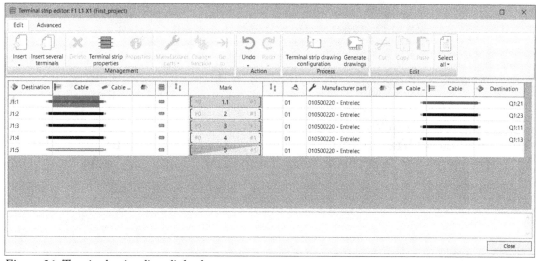

Figure-36. Terminal strip editor dialog box

- From this dialog box, we can find out how a component is connected to a terminal. From the above figure, we can find out that first point of J1 component is connected to first terminal of terminal strip. But, we cannot find out the type of cable used to this connection and manufacturer data for terminal. These parameters are specified in this dialog box.
- To specify the cable type, right-click in the cell under **Cable** column in the dialog box. A shortcut menu will be displayed; refer to Figure-37.
- Click on the **Associate cable cores** option from the shortcut menu. The **Associate cable cores** dialog box will be displayed; refer to Figure-38.

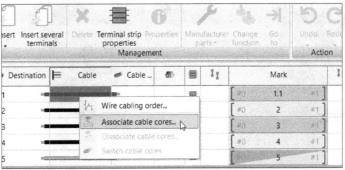

Figure-37. Shortcut menu for terminal cables

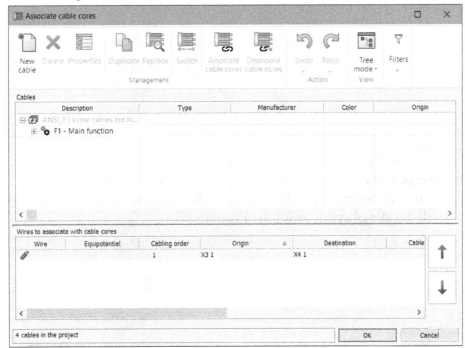

Figure-38. Associate cable cores dialog box

- Click on the **New cable** button from the **Management** panel in the dialog box. The **Cable references selection** dialog box will be displayed; refer to Figure-39.

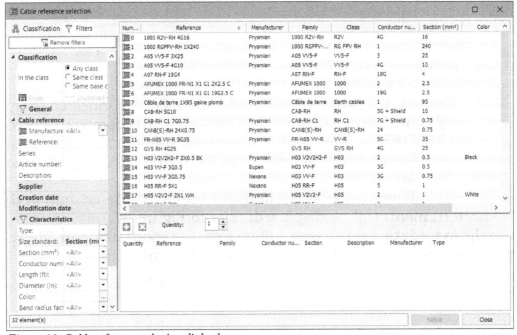

Figure-39. Cable references selection dialog box

- Double-click on the 10 AWG wire from Lapp manufacturer by using filters and click on the **Select** button. The wire will be added in the **Associate cable cores** dialog box.
- Select the wire and click on the **Associate cable cores** button from the dialog box; refer to Figure-40. The wire will be associated with selected cable.

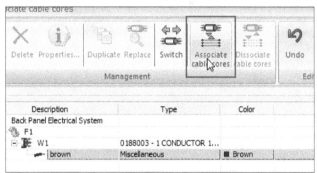

Figure-40. Associate cable cores button

- Similarly, you can associate other wires to cables. Note that if you need the same wire for next association then you can duplicate the cable by using **Duplicate** button after selecting cable from the **Associate cable cores** dialog box; refer to Figure-41.

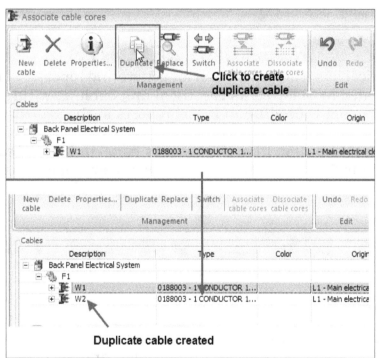

Figure-41. Creating duplicate cable

- To define manufacturer part for a terminal in terminal strip, click in the field under **Manufacturer** column and click on the **Assign manufacturer parts** option from the **Manufacturer parts** drop-down in the **Terminal strip editor** dialog box; refer to Figure-42. The **Manufacturer part selection** dialog box will be displayed.

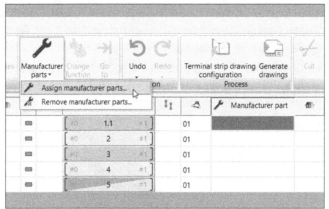

Figure-42. Assign manufacturer parts option

- Select desired manufacturer part for the terminal. Similarly, select the manufacturer part for other terminals.

Note that if you have already associated cables in the schematic drawing then you do not need to specify the cable association again.

- After specifying desired parameters, click on the **Close** (**X**) button at top-right corner of the dialog box to exit.

INSERTING REPORTS

SolidWorks Electrical has a dedicated tool for generating reports. Some of the reports that can be generated in SolidWorks Electrical are; cabling, wiring, Bill of Material, Drawing list, and so on. Procedure to insert reports in the project is given next.

- Click on the **Reports** button from the **Reports** panel in **Electrical Project** tab of the **Ribbon**; refer to Figure-43. The **Report management** dialog box will be displayed as shown in Figure-44.

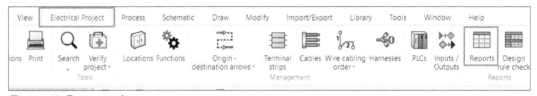

Figure-43. Reports tool

- By default, four reports are displayed in the left of the **Report management** dialog box for Bill of Materials, list of wires, list of cables, and list of drawings. To display the content of the report, click on it from the left of **Report management** dialog box.
- To generate drawing from the report, click on the report from the left of **Report management** dialog box and then click on the **Generate drawings** button from **Report management** dialog box; refer to Figure-45. The drawing will be added in the project; refer to Figure-46.

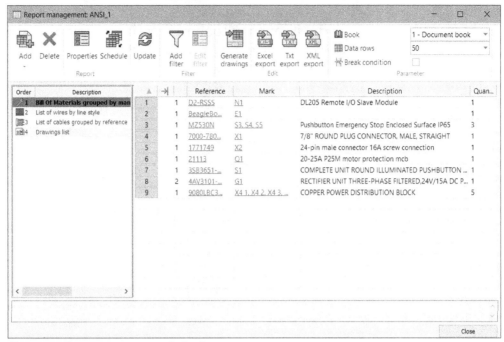

Figure-44. Report management dialog box

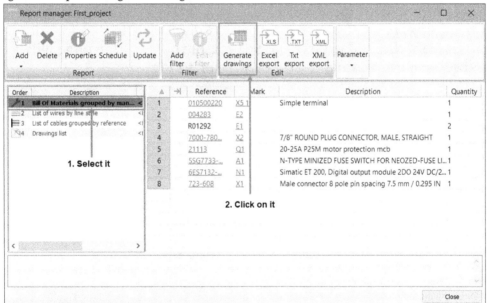

Figure-45. Generating drawing from report

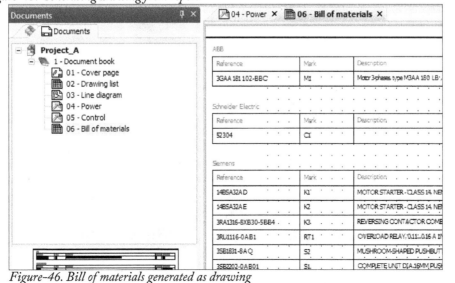

Figure-46. Bill of materials generated as drawing

- You can export the report to external formats like Excel, Txt, and XML by using **Excel export**, **Txt export**, and **XML export** tool, respectively. To do so, click on the respective button (Excel export in our case), the export wizard will be displayed; refer to Figure-47.

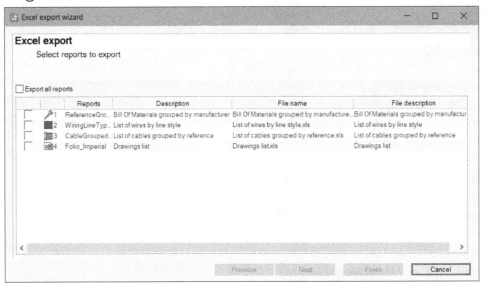

Figure-47. Excel export wizard dialog box

- Select check box/boxes from the list which you want to export and click on the **Next** button from the dialog box. The **Select output folder** page will be displayed in the **Excel export wizard** dialog box; refer to Figure-48.

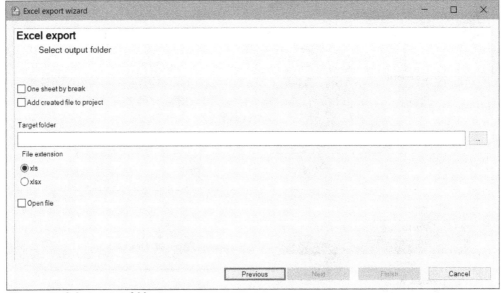

Figure-48. Select output folder page

- Select the **One sheet by break** check box if you want to write all the reports in one sheet separated by breaks.
- Select the **Add created files to project** check box to add the exported files into project also.
- Specify the location of exported file in **Target folder** edit box. You can use the ellipse(...) button next to edit box for selecting folder.
- Select desired format for file from the **File extension** area.
- Select the **Open file** check box to open the file in default application after exporting.
- After setting desired parameters, click on the **Finish** button to export files.

Adding New Report Types

- Click on the **Add** button from the **Report** panel in the **Report management** dialog box. The **Report configuration selector** dialog box will be displayed; refer to Figure-49.

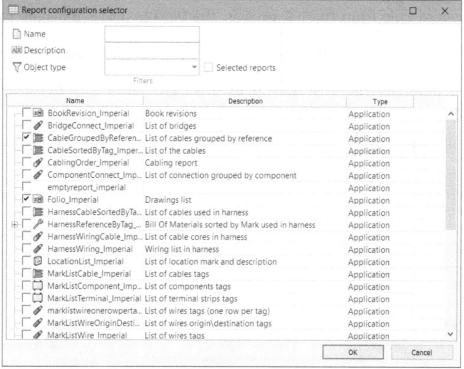

Figure-49. Report configuration selector dialog box

- Select the check boxes for report types to be added and click on the **OK** button. The new report types will be added in the **Report management** dialog box.

PERFORMING DESIGN RULE CHECK

The **Design rule check** tool is used to perform some simple calculations based on specified template. The procedure to use this tool is given next.

- Click on the **Design rule check** tool from the **Reports** panel in the **Electrical Project** tab of the **Ribbon** after creating the drawings. The **Design rules management** window will be displayed; refer to Figure-50.

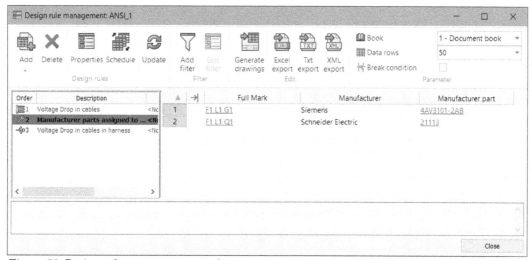

Figure-50. Design rules management window

- Click on the **Add** button from the **Add** drop-down in the **Design rules** panel in the **Ribbon** of window to add a new template for checking results. The **Design rule configuration selector** dialog box will be displayed; refer to Figure-51.
- Select check boxes for the design rules to be analyzed and click on the **OK** button from the dialog box.
- Select desired rule from the left area in the **Design rule manager** to check the results.
- You can export the results in various formats as discussed earlier for report generation.
- Close the manager using **x** button at top right corner or **Close** button after performing design rule checks.

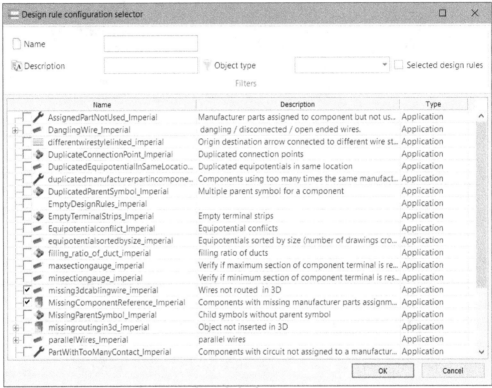

Figure-51. Design rule configuration selector dialog box

Chapter 5

Wire Numbering
and Customizing

The major topics covered in this chapter are:

- *Wire Numbering*
- *Harness*
- *Cable Management*
- *Wire Cabling Order*
- *PLCs Manager*
- *PLC Input/Output Manager*
- *Title Block Manager*
- *2D Footprint Manager*
- *Symbols Manager*
- *Cable references manager*
- *Manufacturer parts manager*

ADDING WIRE NUMBERS MANUALLY

Wiring numbers are used to identify the wires and their connections. The procedure to display wire numbers is given next.

- Click on the **Number new wires** tool from the **Processes** panel in the **Process** tab of the **Ribbon**. The **SOLIDWORKS Electrical** dialog box will be displayed asking whether you want to number new wires or not.
- Click on the **Yes** button from the dialog box. The wiring numbers will be displayed; refer to Figure-1. (Wire numbers are in red boxes).

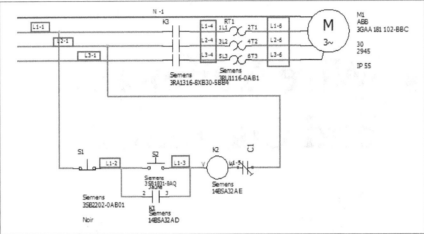

Figure-1. Wiring numbers linked to wires

RENUMBERING WIRES

Once you have added new wires or modified any wire number in the drawing, then you can use the **Renumber wires** tool to modify all the wire numbers.

- Click on the **Renumber wires** tool from the **Processes** panel in the **Process** tab of **Ribbon** if you have made changes in the wiring/created new wires and want to reflect it in the circuit diagram. The **Renumber wires** dialog box will be displayed; refer to Figure-2.
- Select desired radio button from the **Selection** section of the dialog box to define range of drawings in which wires will be renumbered. Select the **Whole project** radio button to change wire numbers of all the drawings in project. Select the **Current book** radio button to change wire numbers of all the drawings in the current book. Select the **Current scheme** radio button change wire numbers in current schematic drawing. Select the **Wires selected in the scheme** radio button to renumber only selected wires. Note that the **Wires selected in the scheme** radio button will be active only if you have selected wires before activating the **Renumber wires** tool. Select the **Choose schemes** radio button to select the schematic drawings in which wires will be renumbered. On selecting this radio button, the **Selection of** dialog box will be displayed; refer to Figure-3. Select desired drawings while holding **CTRL** key and click on the **OK** button.
- Select the **Remove wire numbers** radio button if you want to remove wire numbers. After selecting radio button, select the **Delete manual numbers** check box to delete all wire numbers including manually specified wire numbers. Select the **Reset wire number position** check box to reset position on wire numbers on the wires.
- Select the **Number new wires and recalculate existing wire marks** radio button to apply wire numbers to new wires and recalculate wire numbers of existing wires.

- Select the **Recalculate the wire numbers** radio button to renumber all the wires in sequence. Select the **Recalculate manual numbers** check box if you want to renumber only manually applied wire numbers.
- After setting desired parameters, click on the **OK** button from the dialog box. The **SOLIDWORKS Electrical** information box will be displayed asking to confirm whether you want to renumber all wires or not.
- Click on the **Yes** button from information box. All the wire numbers will be updated.

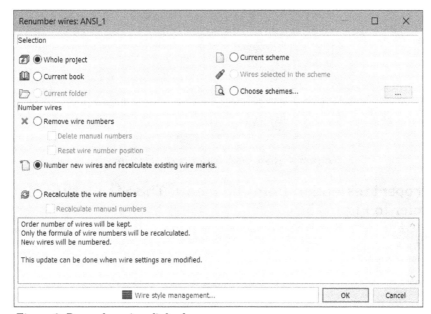

Figure-2. Renumber wires dialog box

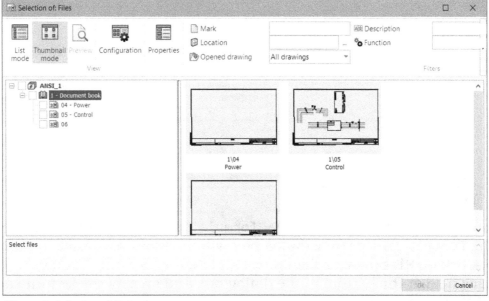

Figure-3. Selection of dialog box

Editing Wire Number Formula

By default, a number is assigned to the wire denoting its wire number but terminals of symbols also have numbers which can sometimes cause confusion. To solve this problem, we can name all wire numbers to be displayed as **Wire n**. Here **n** is the number of wire. Let's see how we can do it.

- Click on the **Wire styles** option tool from the **Configurations** drop-down of the **Project** panel in the **Project** tab of the **Ribbon**. The **Wire style manager** will be displayed as discussed earlier.
- Select the wire style whose wire numbering formula is to be changed and right-click on it. A shortcut menu will be displayed; refer to Figure-4.

Figure-4. Right-click shortcut menu for wire

- Select the **Properties** option from the menu. The **Wire style** dialog box will be displayed; refer to Figure-5.

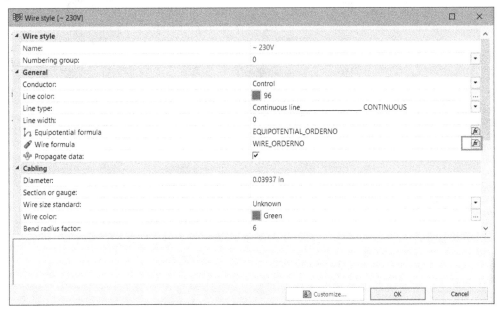

Figure-5. Wire style dialog box

- Click on the **fx** button next to **Wire formula** field in the **General** node of dialog box. The **Formula Management** dialog box will be displayed; refer to Figure-6.
- Double-click on **Wire number** field if not selected by default. Click in the **Formula: Wire mark** field at the bottom in the dialog box and specify the formula as **"Wire"+ WIRE ORDERNO**; refer to Figure-7.

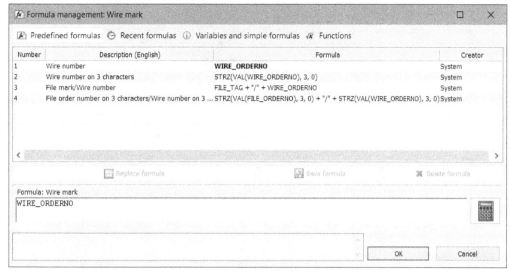

Figure-6. Formula manager

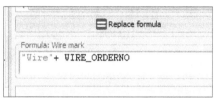

Figure-7. Formula specified in the bottom field

- Click on the **OK** button from the dialog box and then click on the **OK** button from the **Wire style** dialog box. Close the **Wire style management** dialog box. Now, check the wire numbering. If it did not change then click on the **Renumber wires** tool from the **Processes** panel of the **Process** tab in the **Ribbon** and click on the **OK** button from the **Renumber wires** dialog box displayed. The wire numbering will be changed based on specified formula; refer to Figure-8.

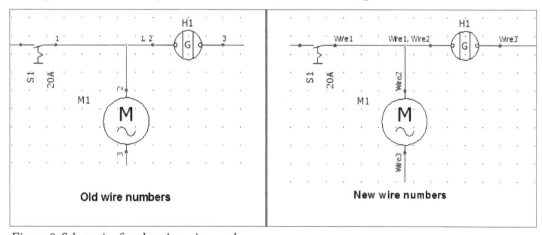

Figure-8. Schematic after changing wire number

HARNESSES

The **Harnesses** tool is used to create harness in electrical project which can later be used to create 3D electrical model. Harness is collection of cables and wires with various connectors. The procedure to create harness is given next.

- Click on the **Harnesses** tool from the **Management** panel in the **Electrical Project** tab of the **Ribbon**. The **Harness management** dialog box will be displayed; refer to Figure-9.

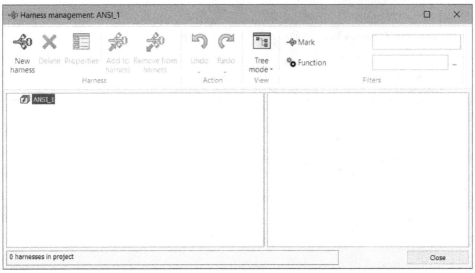

Figure-9. Harness manager

- Click on the **New harness** tool from the **Harness** panel in the **Harness management** dialog box. The **Harness properties** dialog box will be displayed; refer to Figure-10.

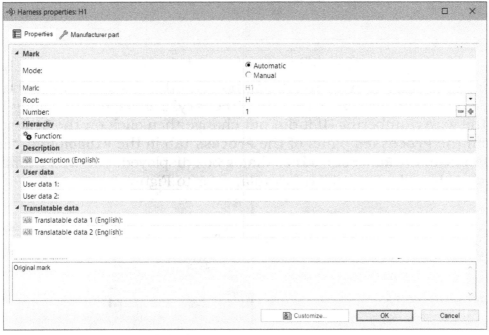

Figure-10. Harness properties dialog box

- Set desired parameters in the **Properties** tab and select desired manufacturer data from **Manufacturer parts** tab. Click on the **OK** button to create harness. A new harness will be created; refer to Figure-11.

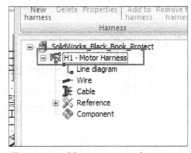

Figure-11. Harness created

- Close the **Harness manager** by clicking on the **Close** button.

- Select all the components and wires that you want to be added in harness from schematic or line diagram, and right-click on them. A shortcut menu will be displayed; refer to Figure-12.

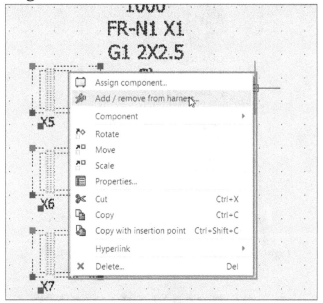

Figure-12. Shortcut menu for harness

- Click on the **Add/remove from harness** tool from the shortcut menu. The **Add/ remove from harness** page of **Command Browser** will be displayed with selected wires and components; refer to Figure-13.

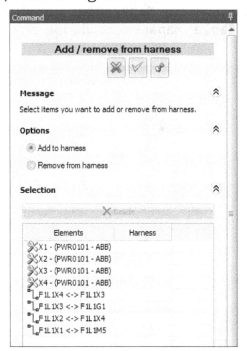

Figure-13. Add remove from harness page in Command Browser

- Make sure the **Add to harness** radio button is selected and then click on the **OK** button from the panel. The **Harness selector** dialog box will be displayed asking you to select a harness.
- Select desired harness from the dialog box and click on the **Select** button from the dialog box. The selected wires and components will be added to the harness. After adding objects to harness, click on the **Close** button to exit **Add/remove from harness** command panel.

If you check the harness in the **Harness manager** then you will find that components and wires are added to harness in respective categories; refer to Figure-14.

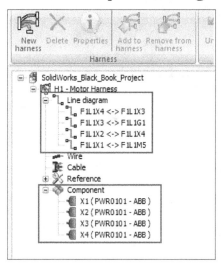

Figure-14. Wires and components added to harness

CABLE MANAGEMENT

You have learned about drawing cable in previous chapters. In this section, you will learn about managing cables. The procedure to do so is given next.

• Click on the **Cables** tool from the **Management** panel in the **Electrical Project** tab of the **Ribbon**. The **Cable management** dialog box will be displayed; refer to Figure-15.

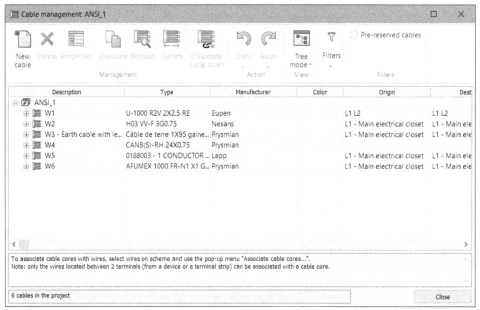

Figure-15. Cables manager

• Click on the **New cable** tool from the **Management** panel in the **Cable management** dialog box. The **Cable references selection** dialog box will be displayed; refer to Figure-16.

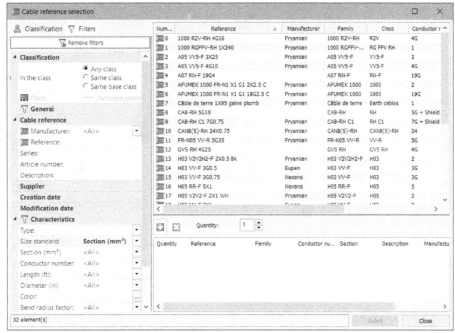

Figure-16. Cable references selection dialog box

- Set desired number of conductors, size, and other parameters in the **Filters** area of the dialog box. The available cables will be displayed in the right area of the dialog box.
- Double-click on the cable(s) that you want to use in your circuit and then click on the **Select** button from the dialog box. The cable(s) will be added in the **Cable management** dialog box; refer to Figure-17.

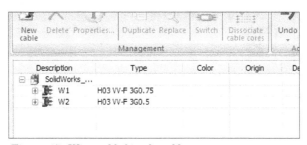

Figure-17. Wires added in the cable manager

- If you want to create more copies of a cable then select the cable from the **Cables management** dialog box and click on the **Duplicate** button from the dialog box. Now, click on **Duplicate** button for the number of times you want the selected cable copies.
- If you want to replace a cable with another cable then select the cable and click on the **Replace** button from the **Management** panel in the dialog box. The **Cable reference selection** dialog box will be displayed. Select desired cable and click on the **Select** button from the dialog box.
- To set the length of cable and other parameters, select the cable in dialog box and click on the **Properties** button from the dialog box. The **Cable** dialog box will be displayed; refer to Figure-18. Set the length, color, voltage-drop, and other parameters of cable in the dialog box and click on the **OK** button. The properties will be applied to cable.
- Now, we have created cables but we have not mentioned what does it connect to. To connect the cables, we need wires drawn in the schematic or block diagram. Close the **Cables manager** by clicking on the **Close** button. Select the wire(s)

that you want to associate with a cable from the diagram and right-click on it. A shortcut menu will be displayed; refer to Figure-19.

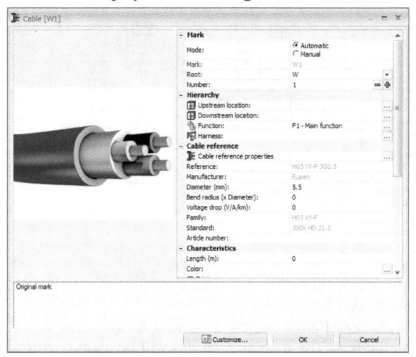

Figure-18. Cable dialog box

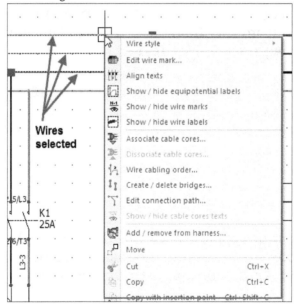

Figure-19. Shortcut menu

- Click on the **Associate cable cores** option from the shortcut menu. The **Associate cable cores** dialog box will be displayed; refer to Figure-20.
- Expand the cable in dialog box to display its cores and select the cores to be associated. Click on the **Associate cable cores** button from the **Management** panel in the dialog box. The cable will be associated with selected wires; refer to Figure-21. Similarly, you can associate other cables. Click on the **OK** button to apply association.

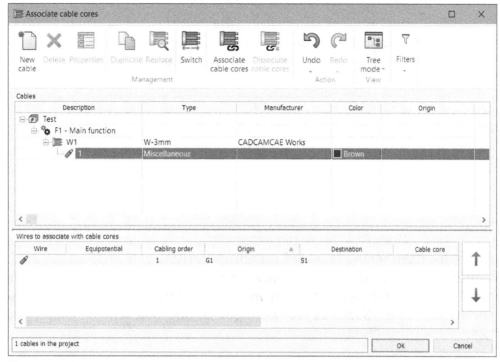

Figure-20. Associate cable cores dialog box

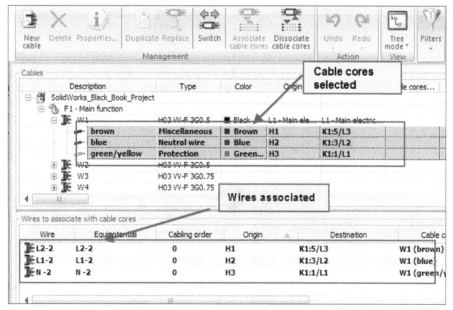

Figure-21. Cable associated with wires

WIRE CABLING ORDER

The **Wire Cabling Order** options are used to change the order by which the wires are connected to components. Take an example of the schematic given in Figure-22. In this figure, wire 1 coming from switch **S1** and wire 2 coming from motor **M** are getting joined at first terminal of glow bulb **H1** named at terminal as **1,2**. What if practically we need the two wires to join at motor terminal and not at glow bulb terminal. The procedure is given next.

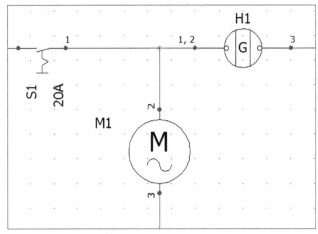

Figure-22. Circuit for wire cabling order

- By default, wire numbering is based on equipotential; refer to Figure-23. To change the wire numbering scheme for displaying individual wire numbers, click on **Wire styles** tool from the **Configuration** drop-down in the **Project** panel of the **Project** tab in the **Ribbon**. The **Wire style manager** dialog box will be displayed; refer to Figure-24.

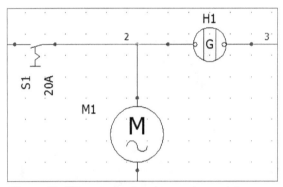

Figure-23. Wire number based on equipotential

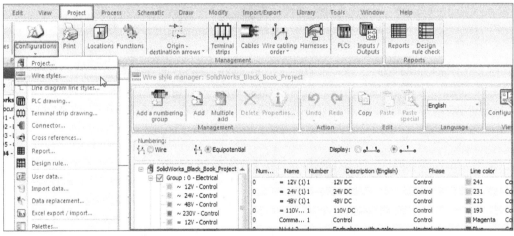

Figure-24. Wire styles tool and Wire style manager

- Select the **Wire** radio button from the **Numbering** area of the dialog box and click on the **Apply** button. The wire numbering scheme will get changed accordingly. Click on the **Close** button. Click on the **Renumber wires** tool and click **OK** from the dialog box if required.
- Click on the **Wire cabling order** tool from the **Wire cabling order** drop-down in the **Management** panel of the **Electrical Project** tab in the **Ribbon**; refer to Figure-25. The **Wire cabling order** dialog box will be displayed; refer to Figure-26.

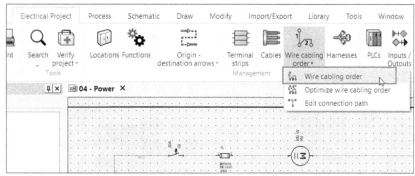

Figure-25. Wire cabling order tool

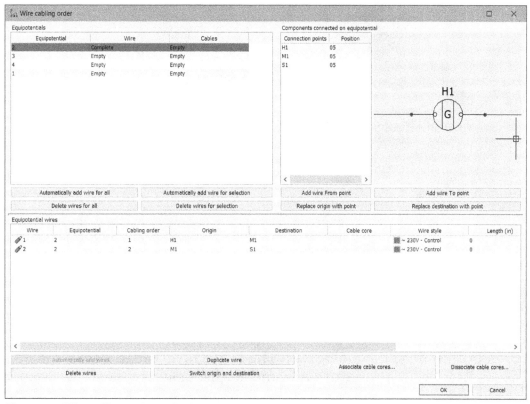

Figure-26. Wire cabling order dialog box

As you can see from above dialog box; the origin of both wires is terminal 1 of H1. But, we want the origin to be 1st terminal of M1. There are two operations required to achieve this: switch origin and destination of wire 2 and change the origin of wire 1. The steps to do so are given next.

- Select the wire **2** from the **Equipotential wires** area of the dialog box. The buttons at the bottom of dialog box will become active.
- Click on the **Switch origin and destination** button from the dialog box. Now, the origin of wire will become **M1:1**.
- Next, we want the origin of wire 1 as M1:1 to do so, click on the **M1:1** connection point from the **Components connected on eqiupotential** area of the dialog box drag-drop it on **H1:1** of wire **1** in **Equipotential wires** area of the dialog box; refer to Figure-27. The origin of wire will get changed; refer to Figure-28.

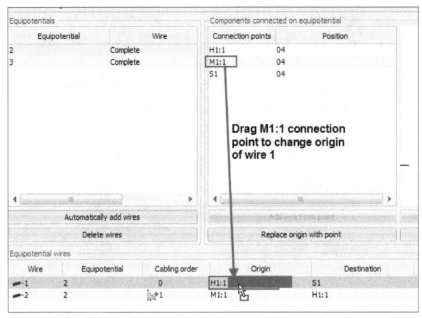

Figure-27. Dragging connection point

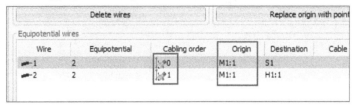

Figure-28. Wires with changed origin

Note that some icons are added in the **Cabling order** column in the **Equipotential wires** area of the dialog box. The icons indicate that the cabling order is changed manually. To set this wiring order as automatic, right-click on the icon and select the **Set as automatic cabling order** option from the shortcut menu. If you want to associate a cable to selected wire then select the wire and click on the **Associate cable cores** button from the dialog box. The **Associate cable cores** dialog box will be displayed. Rest of the procedure is same as discussed earlier.

Optimizing Wire Cabling Order

The **Optimize wire cabling order** tool is used to automatically optimize the wire cabling orders. The procedure to use this tool is given next.

- Click on the **Optimize wire cabling order** tool from the **Wire cabling order** drop-down in the **Management** panel of **Electrical Project** tab in the **Ribbon**. The **Optimize wire cabling order** dialog box will be displayed; refer to Figure-29.
- Select desired radio button from the **Selection** area to define the scope for optimizing wire cabling order. Select the **Whole project** radio button to optimize all wire cabling order in the project. Select the **Current book** radio button to optimize order of all wires in the current book. Select the **Selected locations** radio button to select locations where order will be optimized. Select the **Wires selected in scheme** radio button if you have selected the wires before activating the tool and want to optimize their cabling order.
- Select the **Remove manual wire cabling order** check box if you want to remove manually modified wire cabling order.
- Select the **Remove cable core association** check box if you want to remove cable cores associated with wires in selected scope.

- Select the **Remove bridges** check box if you want to remove bridges assigned to wires in **Terminal strip editor** dialog box.
- Select desired radio button from the **Cabling direction in 2D cabinet layout** area to define the order in which cabling order will be generated.
- After setting desired parameters, click on the **OK** button from the dialog box.

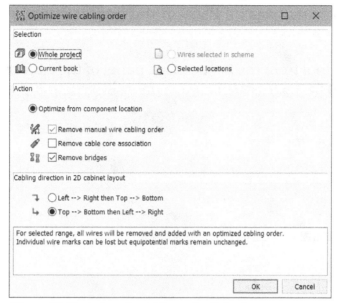

Figure-29. Optimize wire cabling order dialog box

Editing Connection Path

The **Edit connection path** tool is used to modify the representation and order of the wires at intersections. The procedure to use this tool is given next.

- Select the wire which you want to modify and click on the **Edit connection path** tool from the **Wire cabling order** drop-down in the **Management** panel of **Project** tab in the **Ribbon**. The **Edit connection path Command Browser** will be displayed; refer to Figure-30.

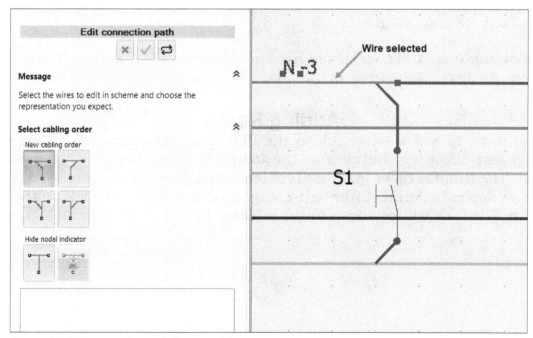

Figure-30. Edit connection path Command Browser

- Select desired button from the **Select cabling order** area of **Command Browser** and click on the **OK** button.
- Click on the **Close** button to exit the **Command Browser**.

PLC MANAGER

PLC Manager is used to manage all PLCs in the project at one place. The procedure to use this tool is given next.

- Click on the **PLCs** tool from the **Management** panel of the **Electrical Project** tab in the **Ribbon**. The **PLCs management** dialog box will be displayed; refer to Figure-31.

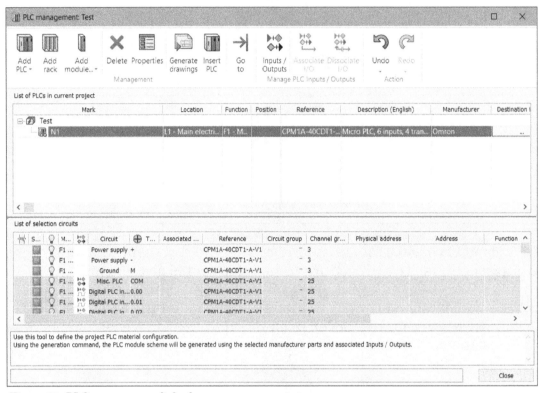

Figure-31. PLC management dialog box

- Click on the **Insert PLC** tool if you have not inserted a PLC earlier. The procedure is same as discussed earlier in the book.

Adding Racks

- If you want to add a new rack to the PLC then select the PLC from the **PLC management** dialog box and click on the **Add rack** button from the **PLC management** panel. The **Manufacturer part selection** dialog box will be displayed. Double-click on desired manufacturer part using the dialog box and click on the **Select** button. The rack will be added to the selected PLC system; refer to Figure-32.

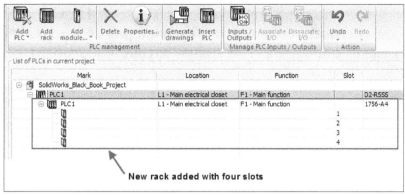

Figure-32. Rack added to PLC

Adding PLC Modules

- If you want to add plc module without interface to the selected rack slot then click on the **Add module** tool from the **Add module** drop-down. If you want to add a plc module with interface then click on the **Add module with interface** tool from the **Add module** drop-down in the **Management** panel of the **PLC management** dialog box. The **Manufacturer part selection** dialog box will be displayed.

- Select desired module using the filters and click on the **Select** button from dialog box. The PLC module will be added to the rack.

Generating PLC Drawings

- If you want to generate drawings for PLCs in **PLC management** dialog box then select desired PLC(s) and click on the **Generate drawings** tool from the **Management** panel in **PLC management** dialog box. The **Selection of: Books, Folders** dialog box will be displayed; refer to Figure-33. Select desired book of the project and click on the **OK** button. The PLC drawings will be added in the project book and **Report creation** box will be displayed. Click on the **Close** button from the **Report creation** box to exit.

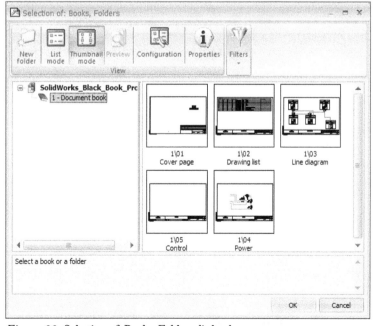

Figure-33. Selection of: Books, Folders dialog box

Managing Inputs/Outputs of PLCs

- Click on the **Inputs/Outputs** tool from the **Manage PLC Inputs/Outputs** panel of the **PLC management** dialog box. The **Input/Output management** dialog box will be displayed; refer to Figure-34.

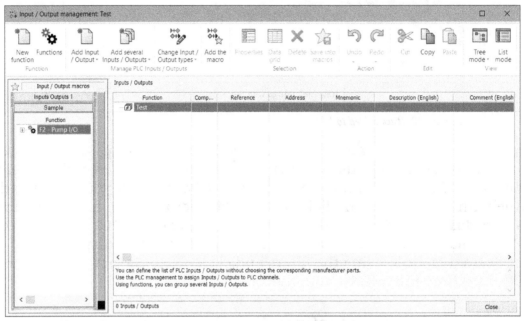

Figure-34. Inputs/Outputs manager

Adding Function

- We divide various inputs & outputs of PLC based on their functions. So, first we need to create functions for input-output. Select the function or book in which you want to add the function and click on the **New function** tool from the **Function** panel in the dialog box. The **Function** dialog box will be displayed; refer to Figure-35.

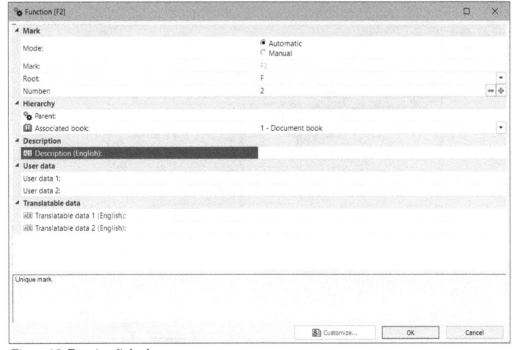

Figure-35. Function dialog box

- Select the **Manual** radio button from **Mark** node and specify desired mark like Motor, Conveyor, and so on in the **Mark** edit box. Click on the **OK** button from the dialog box. The function will be added. To check newly added function, click on the **Functions** tool from the **Ribbon** in **Input/Output management** dialog box. The **Function management** dialog box will be displayed; refer to Figure-36.

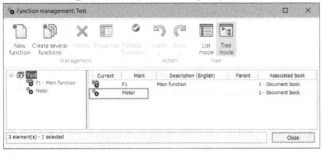

Figure-36. New function added

Adding Input/Output

- Select the function to which you want to add input/output circuits and click on the **Add Input/Output** button from **Manage PLC Inputs/Outputs** panel of the dialog box. A drop-down with input-output options will be displayed; refer to Figure-37.

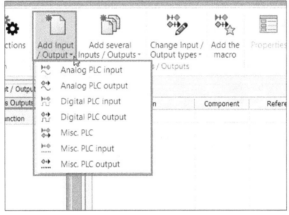

Figure-37. Drop-down with Input/Output options

- Select desired option from the drop-down. The respective input/output will be added to the selected function; refer to Figure-38.

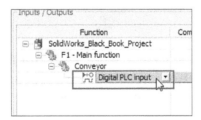

Figure-38. Digital PLC input added

- If you want to add multiple inputs/outputs then click on **Add several Inputs/ Outputs** button from the **Manage PLC Inputs/Outputs** panel of the dialog box. A drop-down similar to the one shown for **Add Input/Output** option will be displayed.
- Select desired option from drop-down (We have selected the **Digital PLC input** option). The **Multiple insertion** dialog box will be displayed; refer to Figure-39.

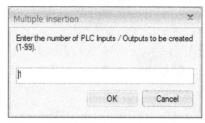

Figure-39. Multiple insertion dialog box

- Specify the number of inputs/outputs to be added in the edit box and click on the **OK** button. The specified number of inputs/outputs will be added. Similarly, add the other inputs/outputs to PLC.
- If you want to edit properties of any of the input or output then select it and click on the **Properties** button from the **Selection** panel in the dialog box. The **Input/ Output properties** dialog box will be displayed; refer to Figure-40. Specify desired properties and click on the **OK** button.

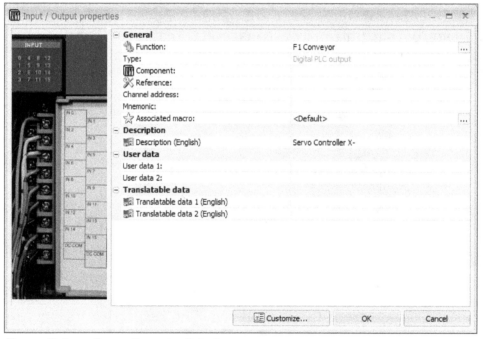

Figure-40. Input Output Properties dialog box

- Close the dialog box once you have specified desired number of inputs and outputs.

Till now we have created PLCs and we have created functions with inputs and outputs, but we have not specified which terminal of PLC does which work. So, we will now connect PLC terminals with Inputs/Outputs specified for PLC.

Connecting PLC Terminals with Input/Output Circuits

- Click again on the **PLCs** tool from the **Management** panel in the **Electrical Project** tab of the **Ribbon** to display **PLC management** dialog box. Select the PLC whose terminals are to be connected. The terminals will be displayed at the bottom in the **PLC management** dialog box; refer to Figure-41.

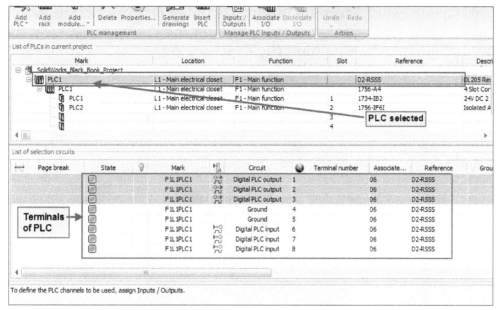

Figure-41. PLC terminals in PLCs manager

- Select the terminals with Digital PLC output circuits and right-click on them. A shortcut menu will be displayed; refer to Figure-42.
- Click on the **Assign an existing PLC Input/Output** option from shortcut menu. The **Inputs/Outputs selection** dialog box will be displayed; refer to Figure-43.

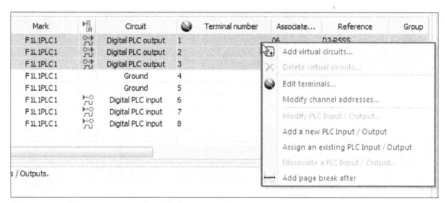

Figure-42. Shortcut menu for PLC terminals

Figure-43. Inputs / Outputs selection dialog box

- Since, we have selected 3 terminals with Digital PLC output circuits in Figure-42 so we need to select 3 Digital PLC outputs from the dialog box. Select the **Outputs** from the dialog box while holding the **CTRL** key and click on the **Select** button. An information box will be displayed confirming the association. Click on the **OK** button.

Similarly, set the other inputs and outputs of PLC terminals. Now, click on the **Generate drawings** tool. If you already inserted PLC drawings then the **Dynamic PLC insertion** dialog box will be displayed; refer to Figure-44.

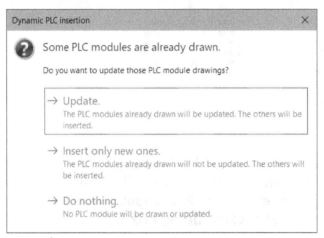

Figure-44. Dynamic PLC insertion dialog box

Click on the **Update** button. Select the book for inserting new drawings and click on the **OK** button from the **Selection of: Books, Folders** dialog box. Close the **Report creation** dialog box and other dialog boxes. The PLC drawings will be added and updated.

TITLE BLOCKS MANAGER

The **Title block management** tool is used to create and manage title blocks. Using this tool, you can create title blocks as per your requirement. The procedure to create title blocks is given next.

- Click on the **Title block management** tool from the **Graphics** panel in the **Library** tab of the **Ribbon**. The **Title block management** dialog box will be displayed; refer to Figure-45.
- To edit any title block template, double-click on it from the **Title blocks manager**. The options for editing block will be displayed; refer to Figure-46.
- Using the sketching tools available in the **Draw** tab of **Ribbon**, you can create the boundary of the template; refer to Figure-47.

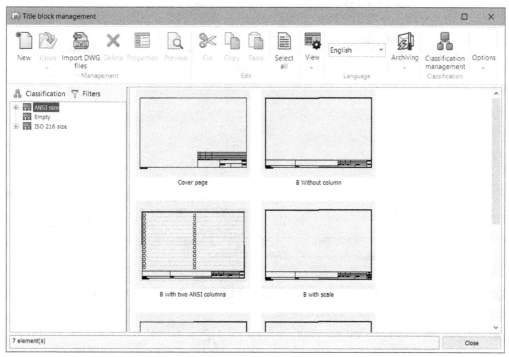

Figure–45. Title blocks manager dialog box

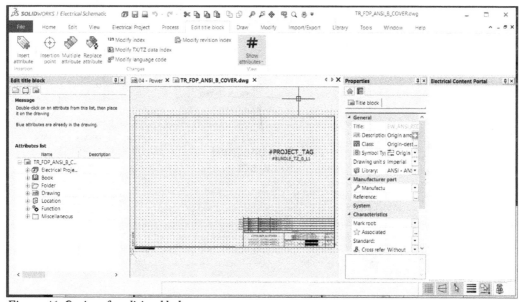

Figure–46. Options for editing block

- Expand the categories in Attributes list and double-click on desired attribute tag to insert it in the title block; refer to Figure-48.

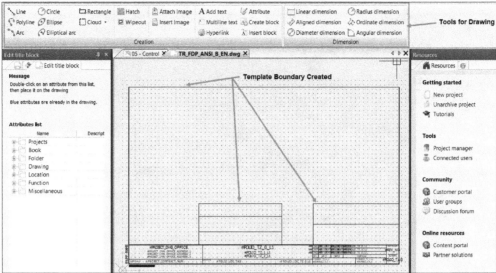

Figure–47. Drawing template boundary

- The tag will be attached to cursor and the **Attribute insertion** Command Browser will be displayed; refer to Figure-49.

Figure–48. Inserting attribute tag

Figure–49. Title block properties

- Specify desired parameters and click in the drawing to place the tag.
- Save the template drawing file and close it.

SYMBOLS MANAGER

The **Symbol management** tool as the name suggests is used to manage symbols of electrical database. Using this tool, you can edit an earlier created symbol or you can create a new symbols as required. The procedure to use this tool is given next.

- Click on the **Symbol management** tool from the **Graphics** panel in the **Library** tab of the **Ribbon**; refer to Figure-50. The **Symbols management** dialog box will be displayed as shown in Figure-51.

Figure-50. Symbols manager tool

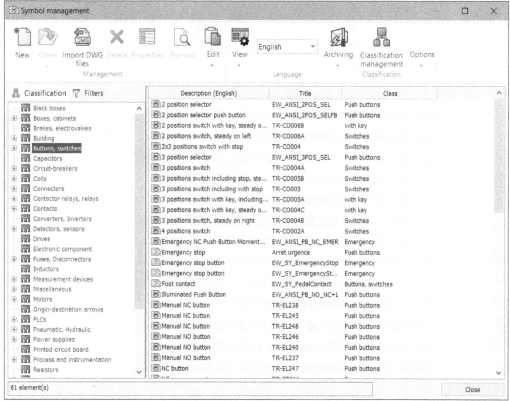

Figure-51. Symbols manager dialog box

- To edit any symbol, double-click on it from the dialog box. Editing environment will be displayed along with the symbol attributes; refer to Figure-52.
- Using the tools in the **Edit symbol** tab in **Ribbon**, you can add connection points, new circuits, attributes, and so on. We will learn about the tools in **Edit symbol** tab later.
- Change the attributes and properties of symbol as done for title block.
- Save the symbol drawing file and close it.

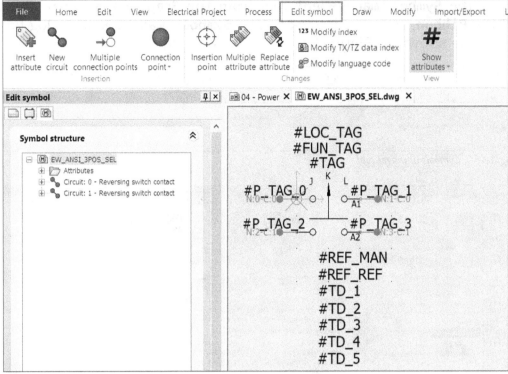

Figure-52. Editing environment for symbol

Creating New Symbol

- To create a new symbol, click on the **New** button from the **Management** panel of **Symbol management** dialog box. The **Symbol properties** dialog box will be displayed; refer to Figure-53.
- Click in the **Symbol name** field and specify desired name for the symbol.
- Click on the Ellipse button [...] for **Class name** field. The **Class selector** dialog box will be displayed; refer to Figure-54.

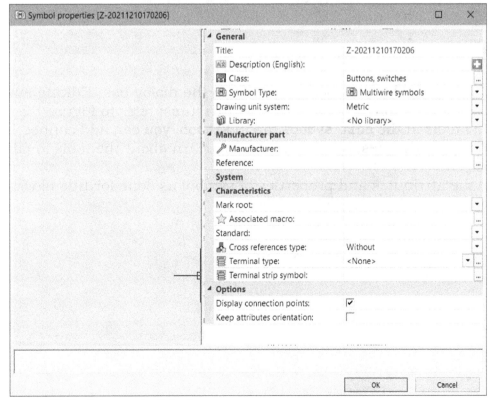

Figure-53. Symbol properties dialog box

- Select desired category for the symbol and click on the **Select** button.
- Click in the **Symbol type** drop-down and select the type of symbol; refer to Figure-55.

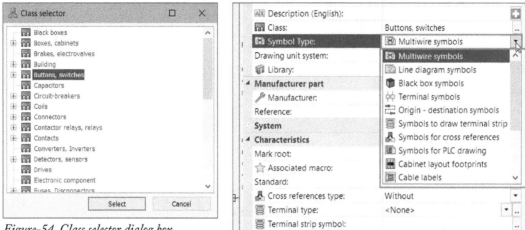

Figure-54. Class selector dialog box

Figure-55. Symbol type drop-down

- Similarly, specify other properties of the symbol and click on the **OK** button from the dialog box. The symbol will be added in the library; refer to Figure-56.

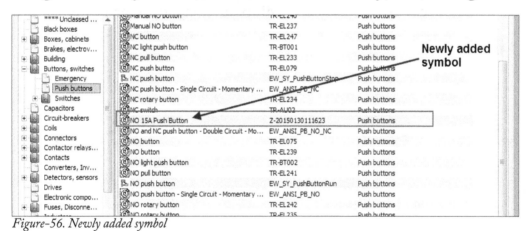

Figure-56. Newly added symbol

- Double-click on the newly added symbol. The symbol editing environment will be displayed with blank drawing area.
- Draw the symbol by using the tools available in the **Draw** tab of **Ribbon**; refer to Figure-57.

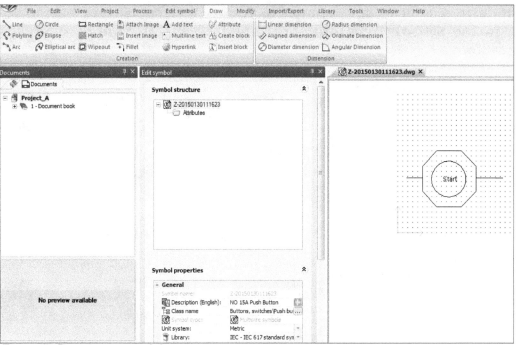

Figure-57. Drawing new symbol

- Click on the **New circuit** tool from the **Insertion** panel in the **Edit symbol** tab of **Ribbon**. The **New circuit** dialog box will be displayed; refer to Figure-58.

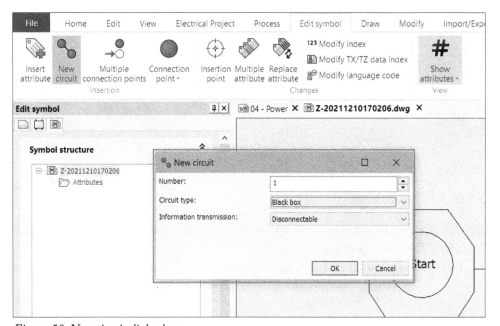

Figure-58. New circuit dialog box

- Specify the number of circuits you want to create for the current symbol by using spinner in the dialog box.
- From the **Circuit type** drop-down, select the circuit (For our case its NO button) and set the transmission information.
- After specifying the settings, click on the **OK** button from the dialog box. The circuit will be added to the symbol structure.
- Click on the down arrow of **Connection point** button in the **Insertion** panel. A list of tools will be displayed; refer to Figure-59.

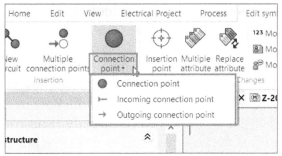

Figure-59. Tools for creating connection points

- The **Incoming connection point** tool is used to set inlet for the component. The **Outgoing connection point** tool is used to set outlet for the component. If the component can be connected by any orientation then the **Connection point** tool is used. Select the circuit from **Symbol structure** section in **Edit symbol Command Browser** at the left in interface and click on desired tool from the drop-down (**Connection point** tool is selected in our case). You are asked to select a reference point on the symbol created.
- Click to specify the connection point. Make sure that you have selected the **OSNAP** button to activate Object snapping for easy point selection; refer to Figure-60.

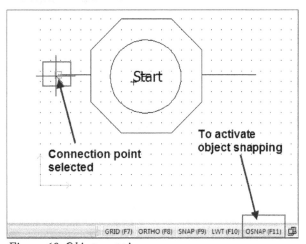

Figure-60. Object snapping

- Similarly, set the connection point on the other side of the symbol. Note that the tag are also attached to the connection points automatically.
- Click on the **Insert attribute** button to insert attributes of symbol. The **Attribute management** dialog box will be displayed; refer to Figure-61.

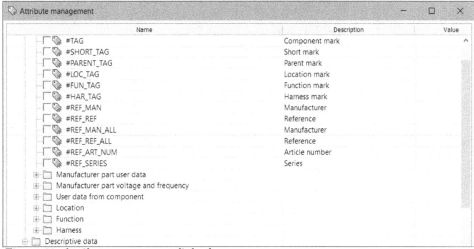

Figure-61. Attribute management dialog box

- To add an attribute to the component, select the respective check box from the dialog box and click on the **OK** button from the dialog box. The tag will get attached to the cursor.
- Click at desired position near the symbol to place the tag.
- Now, click on the **Insertion point** tool from the **Changes** panel in the **Edit symbol** tab to specify the insertion point for the symbol. You are asked to select a point.
- Click at desired location on the symbol to specify it as insertion point. Generally, the best location is one of the connection points earlier specified.
- Now, save the drawing file of symbols and close it. You are ready to use this symbol in drawings.

2D FOOTPRINTS MANAGER

2D footprints are used in designing the used control panel. The 2D footprints manager is used to create and manage footprints. The procedure to use the 2D footprints manager is discussed next.

- Click on the **2D footprint management** tool from the **Graphics** panel in the **Library** tab of **Ribbon**. The **Cabinet layout footprint management** dialog box will be displayed; refer to Figure-62. Here, we will discuss the procedure to create new footprint. You can use the same parameters to edit the footprint.

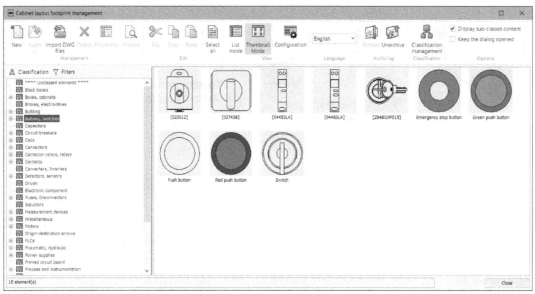

Figure-62. Cabinet layout footprint management dialog box

- Select the category from the **Classification** pane in the left of dialog box to which our new footprint belongs to. Like, we have selected Electronic component category in our case.
- Click on the **New** button from the **Management** panel in the dialog box. The **Symbol properties** dialog box will be displayed as discussed earlier.
- Set desired parameters and click on the **OK** button from the dialog box. The newly added component will be added in the dialog box; refer to Figure-63.

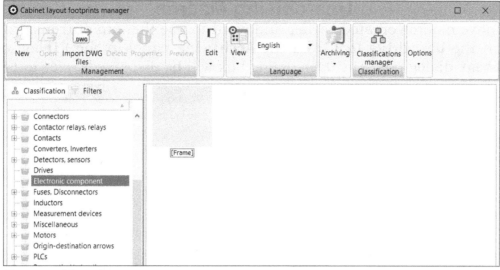

Figure-63. New cabinet component added

- Double-click on the new component and create desired drawing of component.
- After creating desired component, save the file, and then close the file.

MACROS MANAGEMENT

The **Macro management** tool is used to display and manage macros. Macros are piece of ready to use circuits. The procedure to use this tool is given next.

- Click on the **Macro management** tool from the **Graphics** panel of **Library** tab in the **Ribbon**. The **Macro management** dialog box will be displayed; refer to Figure-64.
- Double-click on the macro you want to edit. The schematic drawing for selected macro will open.
- You can modify the drawing as desired using the tools in the **Schematic** tab of **Ribbon**. After modifying the drawing, save it. The macro will be modified accordingly in the **Macros management** dialog box.

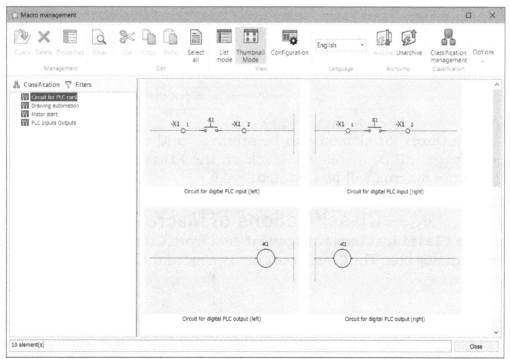

Figure-64. Macro management dialog box

Unarchiving Macros

- If you have macros created earlier and want to use them in current project then click on the **Unarchive** tool from the **Archiving** panel in the **Ribbon** of **Macros management** dialog box. The **Open** dialog box will be displayed; refer to Figure-65.

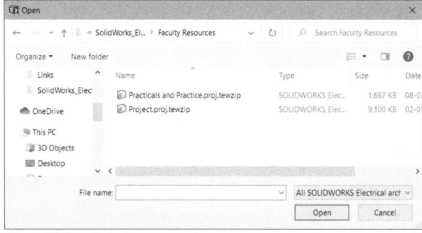

Figure-65. Open dialog box

- Select desired project file which contains macros and click on the **Open** button. The **Unarchiving: Macros** dialog box will be displayed.
- Click on the **Next** button from the dialog box. The **Selection** page of the dialog box will be displayed; refer to Figure-66.

Figure-66. Selection page of Unarchiving dialog box

- Select check boxes for elements to be archived and click on the **Finish** button. The report page will be displayed. Click on the **Finish** button to exit the dialog box. The list of macros will be updated.

Classifications of Macros

- Click on the **Classifications management** tool from **Classification** panel of **Macro management** dialog box. The **Classification management** dialog box will be displayed; refer to Figure-67.

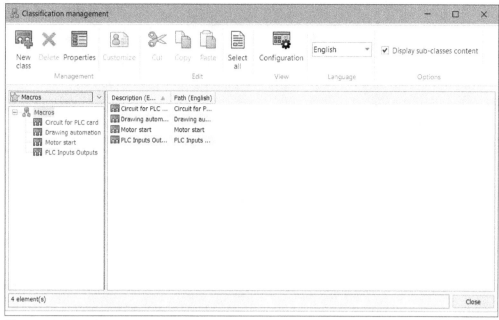

Figure-67. Classification management dialog box

- Click on the **New class** tool from the **Management** panel in the dialog box. The **Class properties** dialog box will be displayed; refer to Figure-68.

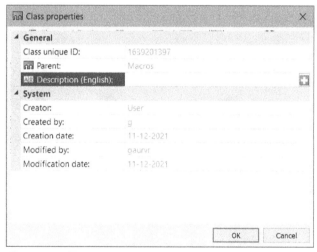

Figure-68. Class properties dialog box

- Specify desired description in the **Description** field and click on the **OK** button. Click on the **Close** button to exit **Classification manager** dialog box.

You can add macros to the newly created classification as discussed earlier.

Inserting Macros

The **Insert macro** tool is used to insert macros earlier created or added in the **Macros manager**. The procedure to use this tool is given next.

- Click on the **Insert macro** tool from the **Insertion** panel in the **Schematic** tab of **Ribbon**. The **Macro selector** dialog box will be displayed; refer to Figure-69.

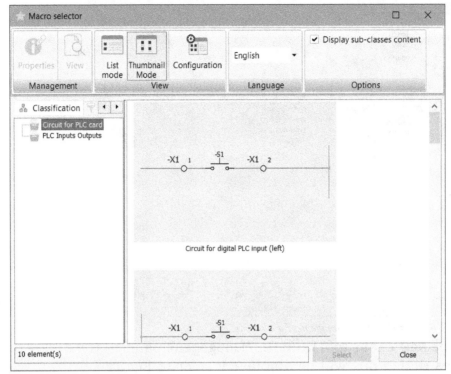

Figure-69. Macro selector dialog box

- Select desired macro and click on the **Select** button. The macro will get attached to cursor.
- Click on the wire or node of component to which you want to connect circuit of current macro. The **Paste special** dialog box will be displayed; refer to Figure-70.

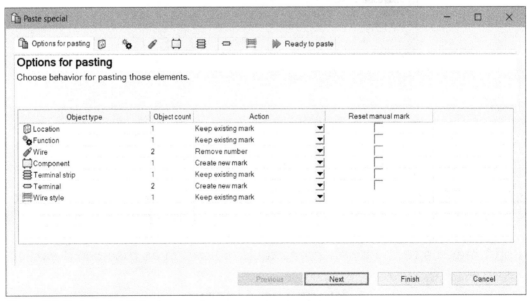

Figure-70. Paste special dialog box

- Set desired options in fields under **Action** column for different object types.
- Click on the **Next** buttons and set desired parameters in the same way.
- After setting desired parameters, click on the **Finish** button from the dialog box to paste the macro.

CABLE REFERENCES MANAGER

The **Cable reference management** tool is used to create and manage cables. You can specify various cable parameters like diameter of wire, length, voltage drop, type of wire, and so on. The procedure to use this tool is given next.

- Click on the **Cable reference management** tool from the **Manufacturers** panel of **Library** tab in the **Ribbon**. The **Cable reference management** dialog box will be displayed; refer to Figure-71.

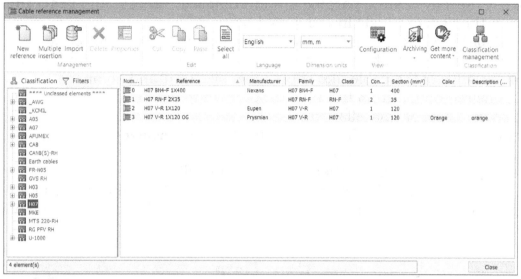

Figure-71. Cable reference management dialog box

- Select desired category from the left. The list of related cables will be displayed on the right in the dialog box.
- Double-click on the cable that you want to edit. The **Cable reference properties** dialog box will be displayed; refer to Figure-72.

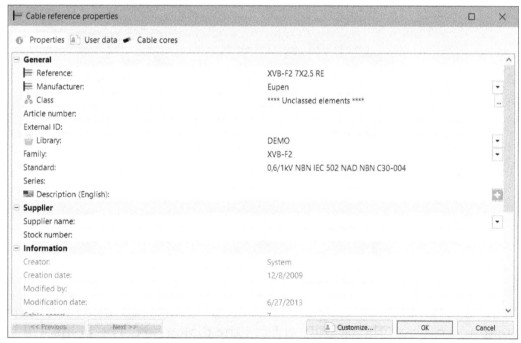

Figure-72. Cable reference properties dialog box

Properties

- Set desired reference code and cable manufacturer data in respective fields.
- Click on the **Browse** button for **Class** field. The **Class selector** dialog box will be displayed. Select desired class for cable and click on the **Select** button.
- Specify desired article number and external id to define manufacturer's data for cable.
- Select the library and family in which you want to place the cable in **Library** and **Family** drop-downs respectively.
- Similarly, specify desired parameters in **Standard**, **Series**, and **Description** edit boxes of the **General** node in the dialog box.
- In the **Supplier** node, specify the name and stock number for cable.
- The options in the **Characteristics** node are used to define functional characteristics of the cable like cable type, size unit and so on.
- Select desired option from the **Type** drop-down to define whether cable is control type, power type, or miscellaneous.
- Select desired unit for cable size from the **Size standard** drop-down like section (mm2), Gauge (AWG standard), and so on.
- Similarly, you can specify other characteristics like conductor section, length of each section of cable, diameter of cable, color of cable, bend radius factor, bend radius, linear mass of each wire/cable section, and voltage drop in volt or ampere per km length.

User Data

The options in the **User Data** tab are used to define user descriptions of the cable. Specify desired text data in the fields of **Description** and **User data** node. If you want to define user data in different language for **Description** and **Translatable data** fields then click on the **+** button next to field. The **Language selector** dialog box will be displayed; refer to Figure-73.

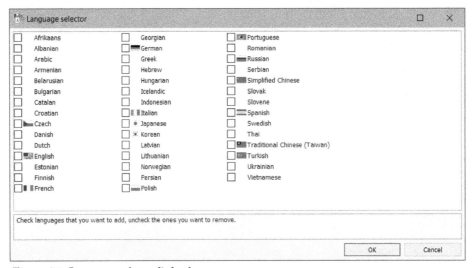

Figure-73. Language selector dialog box

- Select the check boxes for languages from the dialog box in which you want to write text and click on the **OK** button from the dialog box. The fields will be added in the dialog box.

Cable cores data

The options in the **Cable cores** tab are used to define data related to wire cores in the cable; refer to Figure-74.

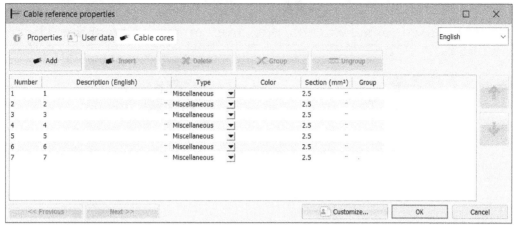

Figure-74. Cable cores tab

- Set desired description, type, color, and section for each cable core.
- Click on the **Add** button from the dialog box. A new cable core will be added in the dialog box. You can add desired number of cores in the cable.
- After setting desired parameters, click on the **OK** button from the dialog box. The **Cable references management** dialog box will be displayed again.

Creating New Cable

The **New reference** tool in **Cable reference management** dialog box is used to create new cable. The procedure to use this tool is given next.

- Click on the **New reference** tool from the **Management** panel of the **Cable reference management** dialog box. The **New cable reference** dialog box will be displayed; refer to Figure-75.

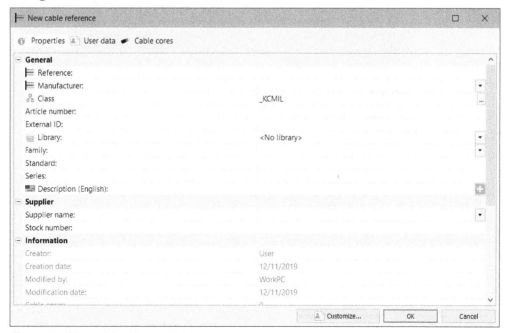

Figure-75. New cable reference dialog box

- Set the parameters as discussed earlier and click on the **OK** button.

If you want to create multiple cables then click on the **Multiple insertion** tool from the dialog box. Rest of the procedure is same as discussed earlier.

Importing Cable Data

The **Import** tool is used to import cable data from an excel or text file. The procedure to use this tool is given next.

- Click on the **Import** tool from the **Management** panel in the **Cable reference management** dialog box. The **Import cable references** dialog box will be displayed; refer to Figure-76.

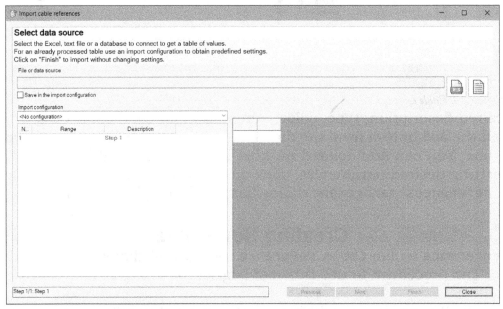

Figure-76. Import cable references dialog box

- Click on the **Excel file** or **Text file** button 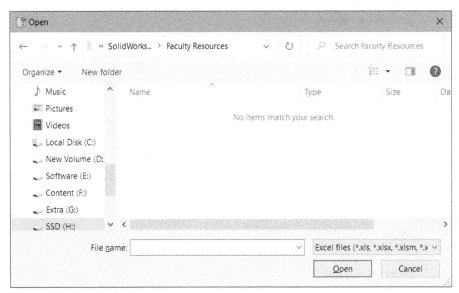 to open respective data source file. The **Open** dialog box will be displayed; refer to Figure-77.

Figure-77. Open dialog box

- Select desired file and click on the **Open** button. The data will be displayed on the right in the dialog box; refer to Figure-78.

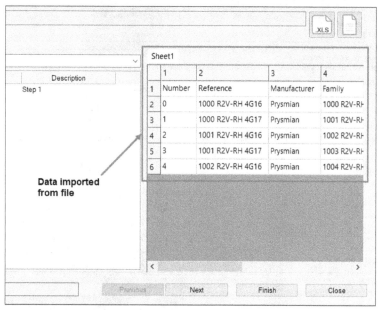

Figure-78. Data imported from excel file

- Select desired option from the **Import configuration** drop-down to define which standard should be used to importing data.
- Click on the **Next** button from the dialog box. The **Definition of a data range** page will be displayed in the dialog box; refer to Figure-79.

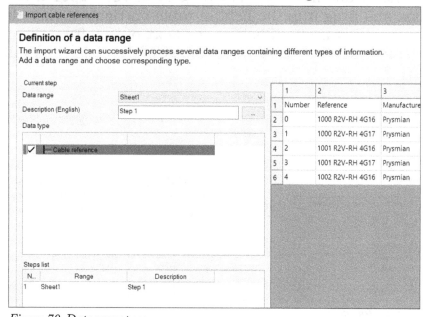

Figure-79. Data range page

- Set desired parameters in the dialog box and click on the **Next** button. The **Title rows** page will be displayed where you can define how many rows of imported data are title rows.
- Set desired value in the **Number of title rows** spinner to define fixed title row. Click on the **Next** button from the dialog box. The **Fields definition** page of dialog box will be displayed; refer to Figure-80.

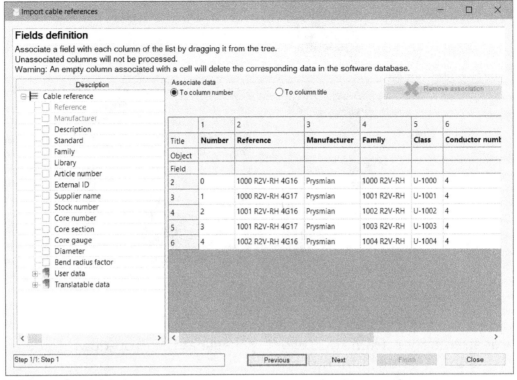

Figure-80. Fields definition page

- Select fields one by one from desired category in **Description** box at the left in the dialog box and drag it to the field in table on right so that the data can be associated; refer to Figure-81. Click on the **Next** button from the dialog box. The **Data comparison - differential report** page of the dialog box will be displayed; refer to Figure-82.

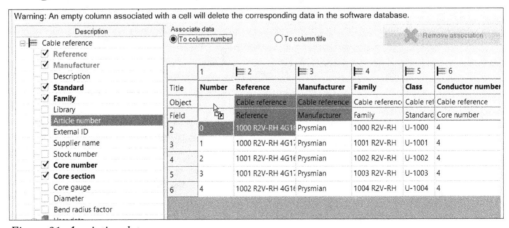

Figure-81. Associating data

- Click on the **Compare** button from the dialog box to check if data is new or there is an update in the data. The report will be generated and displayed; refer to Figure-83.

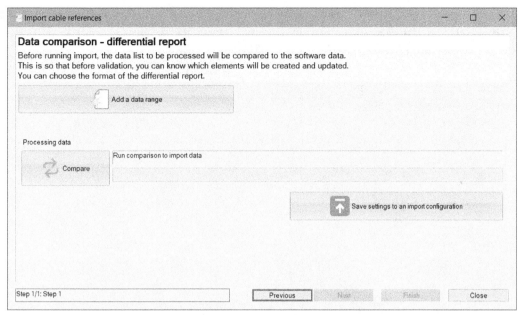

Figure-82. Data comparison-differential report page

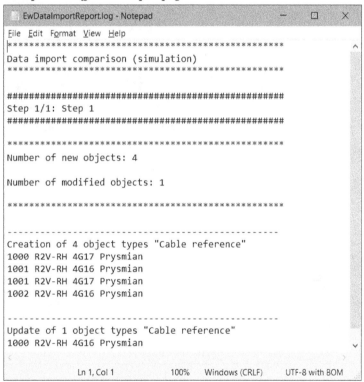

Figure-83. Data import comparison report

- Check the report and close it. Click on the **Import** button from the **Processing Data** area of the dialog box. The data will be imported in database.
- Click on the **Finish** button to complete importing data and close the dialog box by clicking **Close** button.

MANUFACTURER PARTS MANAGER

The **Manufacturer part management** tool is used to create, manage, and import manufacturer data. The procedure to use this tool is given next.

- Click on the **Manufacturer part management** tool from the **Manufacturers** panel of **Library** tab in the **Ribbon**. The **Manufacturer part management** dialog box will be displayed; refer to Figure-84.
- Select desired category from the left and click on the **Search** button. The list of manufacturer parts will be displayed.

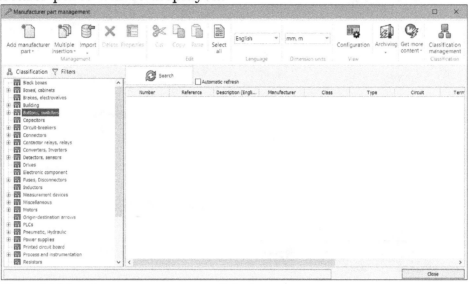

Figure-84. Manufacturer part management dialog box

Modifying a Part

- Double-click on the manufacturer part that you want to modify. The **Manufacturer part properties** dialog box will be displayed; refer to Figure-85.

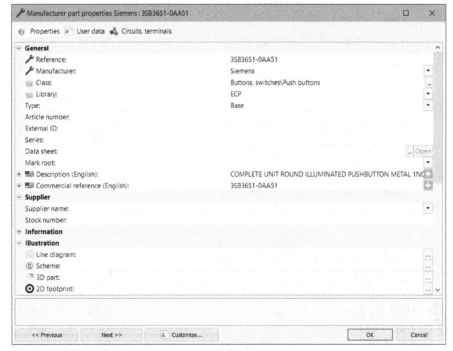

Figure-85. Manufacturer part properties dialog box

- Specify the general data and supplier data of part in the **General** node and **Supplier** node of **Properties** tab as discussed earlier.
- In the **Illustration** node, you can connect line diagram symbol, schematic symbol, 3D SolidWorks part, 2D footprint, connection label, and printed circuit board file.
- Click on the **Browse** button for desired category from the Illustration node like if you want to assign a schematic symbol to the manufacturer part then click on

the **Browse** button for it. The **Symbol selector** dialog box will be displayed; refer to Figure-86.

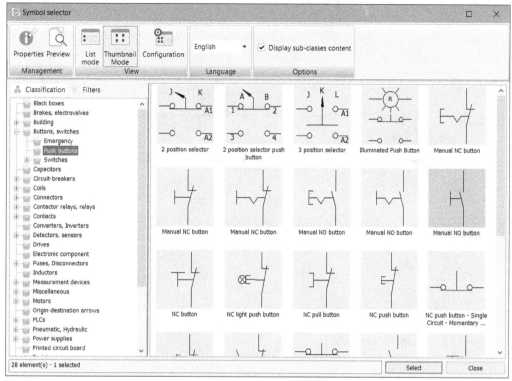

Figure-86. Symbol selector dialog box

- Select desired symbol from the dialog box and click on the **Select** button. The selected symbol will get attached to the manufacturer part.
- Specify desired physical size of manufacturer part in the fields of Size node. This data is useful for fabrication of control panel.
- Specify desired voltage and frequency values in **Use** and **Control** nodes. These values define the electrical data for manufacturer part.
- You can specify the manufacturer data in fields of **Manufacturer data** node in the same way.
- Specify the user data for manufacturer part in the **User data** tab of dialog box.

Defining Circuits and Terminals of part

Click on the **Circuits, terminals** tab of the dialog box. The options will be displayed as shown in Figure-87.

- Click on the **Add** button from the **Circuits** area of the dialog box to add a new circuit. A new circuit will be added in the list.
- Click in the field under **Type** column and select desired type of circuit; refer to Figure-88.
- Note that terminals will be displayed for created new circuit in the **Terminals** area of the dialog box.
- Click on the **Add** button from the **Terminals** area of the dialog box. A new terminal will be added in the list.

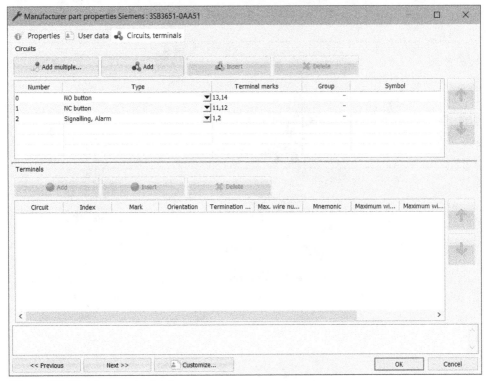

Figure-87. Circuits, terminals tab

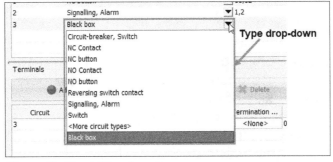

Figure-88. Circuit Type drop-down

- Set desired orientation for the terminal from the **Orientation** field. You can define the orientation as **Incoming**, **Outgoing**, or **Undefined**.
- Click on the field in **Wire termination type** column of the **Terminals** area. The **Wire termination type selector** dialog box will be displayed; refer to Figure-89.
- Select desired terminal and click on the **Select** button.
- Specify the other parameters as desired in the table. Note that values specified in the **Mark** column will be displayed on the part after inserting them in drawing. We generally specify **+** and **-** marks to define polarity of the part.
- After setting desired parameters, click on the **OK** button from the dialog box. The **Manufacturer parts manager** dialog box will be displayed.

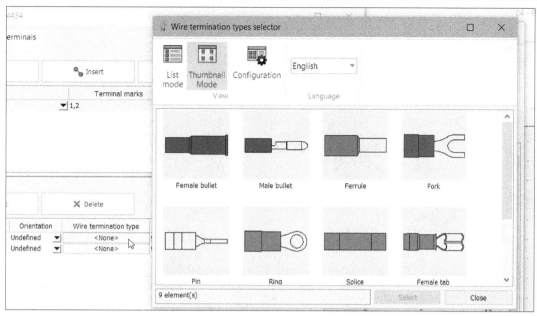

Figure-89. Wire termination type selector dialog box

Creating a New Part

The **Add manufacturer part** tool is used to add new manufacturer part details in the database. The procedure to use this tool is given next.

• Click on the **Add manufacturer part** tool from the **Management** panel in the **Manufacturer parts manager** dialog box. The **Manufacturer part properties** dialog box will be displayed.
• Specify the parameters as discussed earlier and click on the **OK** button.

Adding an Electrical Assembly

An electrical assembly is an assembly of various electrical components. The procedure to add electrical assembly is given next.

• Click on the **Add electrical assembly** tool from the **Add manufacturer part** drop-down in the **Management** panel of **Manufacturer part management** dialog box. The **Electrical assembly properties** dialog box will be displayed; refer to Figure-90.
• The options in this dialog box are same as discussed for **Manufacturer part properties** dialog box.
• Set desired parameters and click on the **OK** button.

The procedure to import manufacturer parts and printed circuit boards is same as discussed earlier. Click on the **OK** button from the **Manufacturer part management** dialog box to exit.

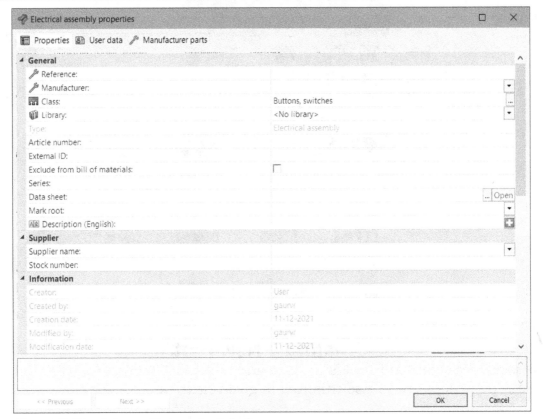

Figure-90. Electrical assembly properties dialog box

WIRE TERMINATION TYPE MANAGER

The **Wire termination type management** tool is used to create and manage terminals for parts. The procedure to use this tool is given next.

• Click on the **Wire termination type management** tool from the **Customization** panel in the **Library** tab of **Ribbon**. The **Wire termination types management** dialog box will be displayed; refer to Figure-91.

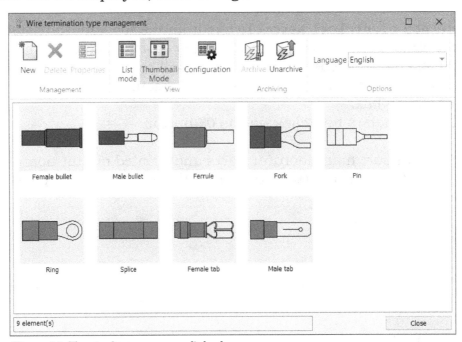

Figure-91. Terminal types manager dialog box

- Double-click on desired terminal that you want to modify. The **Wire termination type properties** dialog box will be displayed; refer to Figure-92.

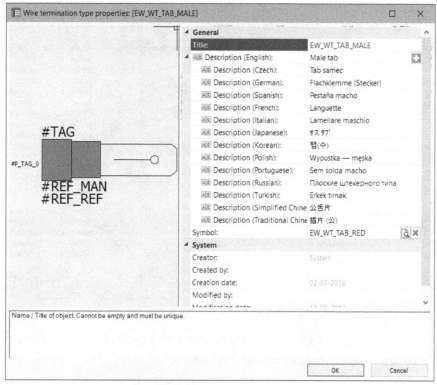

Figure-92. Wire termination type properties dialog box

- Set desired parameters in the dialog box and click on the **OK** button.

Creating New Termination Type

- Click on the **New** button from the **Management** panel in the **Wire termination type management** dialog box. The **Wire termination type properties** dialog box will be displayed.
- Specify desired parameters in the dialog box. Click on the **Search** button for **Symbol** field. The **Symbol selector** dialog box will be displayed.
- Select desired symbol from the dialog box and click on the **Select** button. The symbol will be assigned to the new termination type.
- After setting desired parameters, click on the **OK** button.

You can use **Archive** and **Unarchive** buttons in the **Wire termination type management** dialog box as discussed earlier. Close the dialog box by clicking on the **Close** button.

LIBRARIES MANAGER

The **Library management** tool is used to add and remove libraries of components from the database. The procedure to use this tool is given next.

- Click on the **Library management** tool from the **Customization** panel in the **Library** tab of the **Ribbon**. The **Library management** dialog box will be displayed; refer to Figure-93.

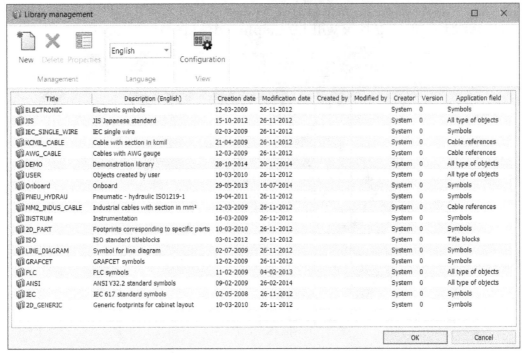

Figure-93. Library management dialog box

- Select desired library and click on the **Properties** button. The **Library properties** dialog box will be displayed; refer to Figure-94.

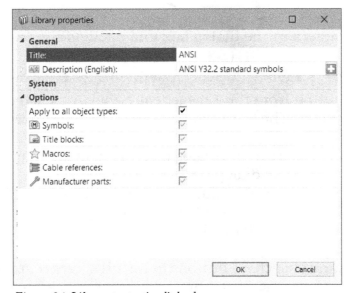

Figure-94. Library properties dialog box

- Select the check boxes for objects to which you want to apply library property. Set desired parameters and click on the **OK** button.
- If you want to delete a library from the database then select it from the dialog box and click on the **Delete** button.
- If you want to add a new library in the database then click on the **New** button from the **Management** panel in the **Libraries manager** dialog box. The **Library properties** dialog box will be displayed. The options of this dialog box have been discussed earlier.
- Click on the **OK** button from the dialog box to apply changes.

CLASSIFICATION MANAGEMENT

The **Classification management** tool is used to create and manage classes of components, cables, macros, and title blocks. The procedure to use this tool is given next.

- Click on the **Classification management** tool from the **Customization** panel of **Library** tab in the **Ribbon**. The **Classification management** dialog box will be displayed; refer to Figure-95.

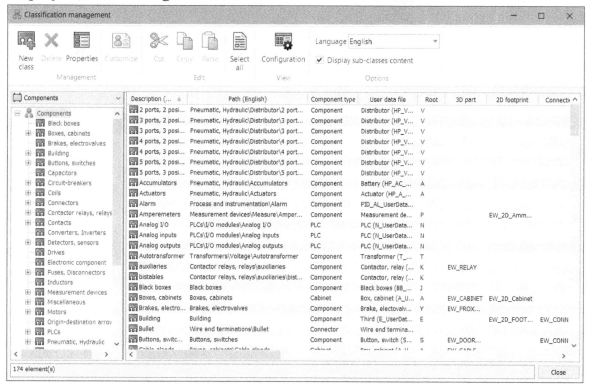

Figure-95. Classification management dialog box

Creating a New Class

- Click on the **New class** tool from the **Management** panel of **Ribbon** in **Classification management** dialog box. The **Class properties** dialog box will be displayed; refer to Figure-96.

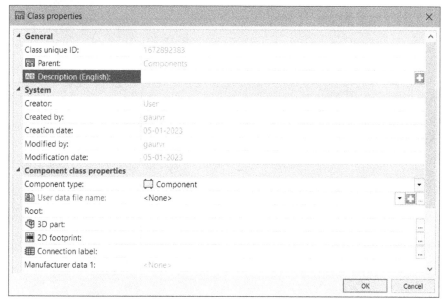

Figure-96. Class properties dialog box

- Specify desired parameters in the dialog box as discussed earlier and click on the **OK** button to create the class.

You can use the other tools in **Classification management** dialog box as discussed earlier. Click on the **Close** button from the dialog box to exit.

TERMINAL TYPE MANAGEMENT

The **Terminal type management** tool is used to create and manage different types of terminals. The procedure to use this tool is given next.

- Click on the **Terminal type management** tool from the **Customization** panel of **Library** tab in the **Ribbon**. The **Terminal type management** dialog box will be displayed; refer to Figure-97.

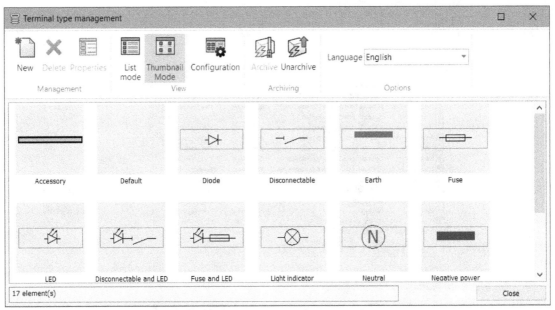

Figure-97. Terminal type management dialog box

- Click on the **New** tool from the **Management** panel of **Ribbon** in dialog box. The **Terminal type properties** dialog box will be displayed; refer to Figure-98.

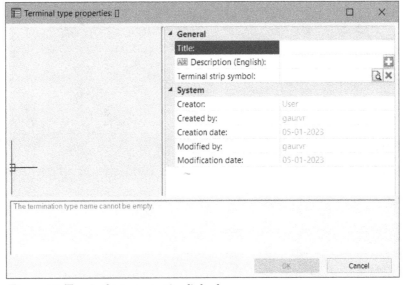

Figure-98. Terminal type properties dialog box

- Set the name, description, and terminal strip symbol for the new terminal type and click on the **OK** button. The new terminal type will be added in the dialog box.
- Click on the **Close** button to exit the dialog box.

PERFORMING ERP DATABASE CONNECTION

The **ERP database connection** tool is used to connect ERP database of your company with SolidWorks Electrical library. The procedure to use this tool is given next.

- Click on the **ERP database connection** tool from the **Customization** panel of **Library** tab in the **Ribbon**. The **Catalogs custom data - ERP connections** dialog box will be displayed; refer to Figure-99.

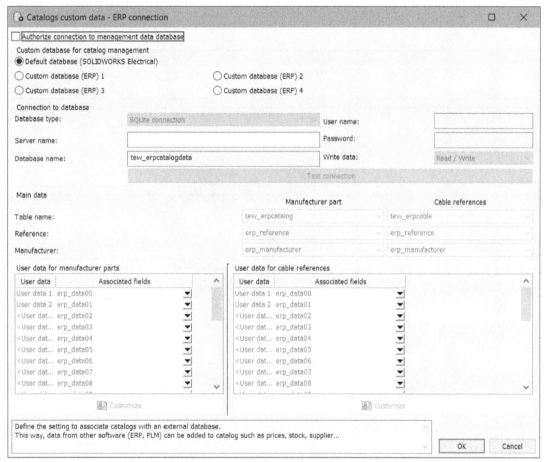

Figure-99. Catalogs custom data-ERP connection dialog box

- Select the **Authorize connection to management data database** tool from the dialog box to activate the connection link between catalogs and external database.
- Select desired **Custom database** radio button to connect to respective database on server.
- Select desired server option from the **Database type** drop-down and specify credentials for accessing database connection.
- Click on the **Test connection** button from the **Connection to database** area of the dialog box.
- Set the other parameters as desired and click on the **OK** button to perform connection.

FOR STUDENT NOTES

Chapter 6

Cabinet Layout

The major topics covered in this chapter are:

- *Introduction*
- *2D Cabinet Layout*
- *Inserting 2D Footprint*
- *Inserting Terminals Strips and Rails*
- *Inserting Ducts*
- *Adding cabinets*
- *Aligning footprints*

INTRODUCTION

In the previous chapters, you have learnt to create schematic circuit diagrams. After creating those circuit diagrams, the next step is to create panels. A panel is the box consisting of various electrical switches and PLCs to control the working of equipment. Refer to Figure-1. Note that the panel shown in the figure is back side panel of a machine. This panel is generally hidden from the operator. What an operator see is different type of panel; refer to Figure-2. We call this panel as User Control panel. In both the cases, the approach of designing is almost same but the interaction with the user is different. The User Control Panel is meant for Users so it can have push buttons, screen, sensors, key board and so on. On the other side, the back panel will be having relays, circuit breakers, sensors, connectors, plcs, switches, and so on for electrical engineer.

Figure-1. Panel

Figure-2. User Control panel

If we start linking the schematic drawings with the panel drawings then the common platform is the component tag and the location code. Suppose we have created a push button in the schematic with tag -04PB2 then in the panel layout you should insert the same push button with the same tag. The location of the Push button will be decided by the Location code. The components that are having same location code should be placed at the same place in the panel. Also, the components that are having the same Function code should be placed together in the cabinet.

There are two options in SolidWorks Electrical to create cabinet layout; **2D cabinet layout** and **SOLIDWORKS assembly**. The **2D cabinet layout** tool is used to create 2 dimensional layout of the cabinet. The **SOLIDWORKS assembly** tool is used to create 3D model of the cabinet. The procedures to use these tools are discussed next.

2D CABINET LAYOUT

The **2D cabinet layout** tool is available in the **Processes** panel of the **Process** tab in the **Ribbon**; refer to Figure-3. This tool is used to generate 2D layout of an electrical panel. The procedure to use this tool is given next.

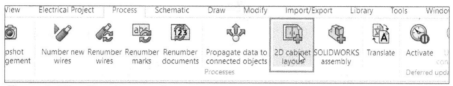
Figure-3. 2D cabinet layout tool

- Click on the **2D cabinet layout** tool from the **Processes** panel. The **Creation of 2D cabinet layout drawings** dialog box will be displayed; refer to Figure-4.

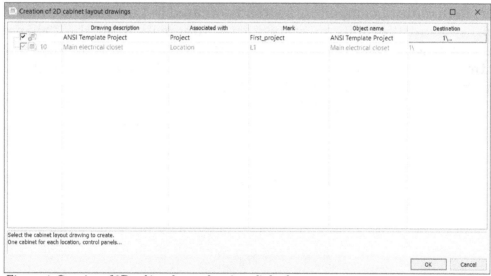
Figure-4. Creation of 2D cabinet layout drawings dialog box

- Select the check box for only those locations for which you want to make the cabinets and then click on the **OK** button. New cabinet drawing/drawings will be added in the **Pages Browser**; refer to Figure-5.

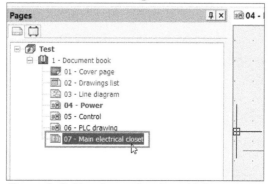
Figure-5. Cabinet layout drawing added

- Double-click on the newly created cabinet drawing. The Cabinet layout editing environment will be displayed; refer to Figure-6.

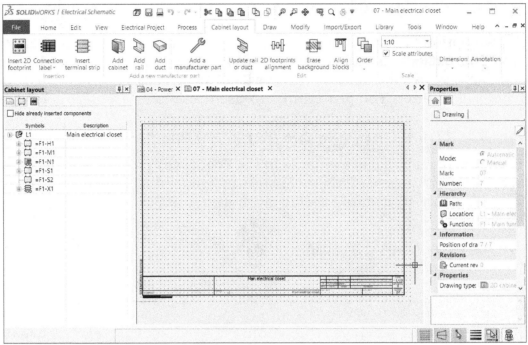

Figure-6. Cabinet layout editing environment

- To insert footprints of components already created in Line diagram and Schematic diagram, expand the node of respective component from the **Cabinet layout browser** and select the corresponding check box; refer to Figure-7. The **2D footprint insertion** page of **Command Browser** will be displayed and the footprint will get attached to the cursor; refer to Figure-8.

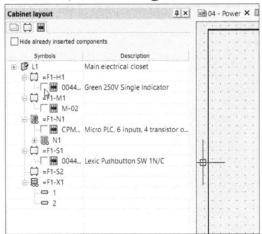

Figure-7. Auto generated footprints for components

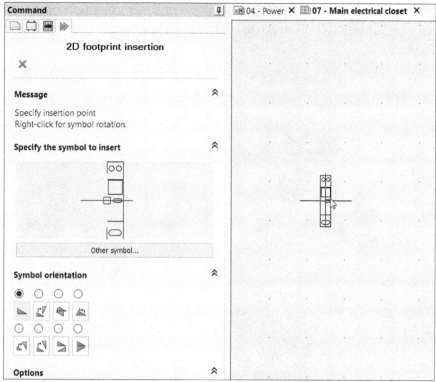

Figure-8. 2D footprint insertion page

- If you want to use another footprint symbol then click on the **Other symbol** button from the **Command Browser** and select desired symbol.
- Click in the drawing to place the footprint. Repeat the procedure to place other footprints.

The tools available in the **Cabinet layout** tab are discussed next.

Insert 2D footprint

This tool is used to insert footprints for the components existing in the Line diagram and Schematic drawing of the project. The procedure to use this tool is given next.

- Select a component from the **Cabinet layout browser** for which you want to place the footprint.
- Click on the **Insert 2D footprint** tool from the **Insertion** panel in the **Cabinet layout** tab of the **Ribbon**. Related footprint will get attached to cursor and the **2D footprint insertion** page will be displayed in the **Command Browser**.
- Click in the drawing area to place the footprint.

The tools available for Connection labels and report table in **Cabinet layout** tab of **Ribbon** work in the same way as discussed earlier.

Insert terminal strip

The terminals are used to connect components with supplies. Use of terminals ensure the safety of wires since in case of short-circuit, wire will get burnt at the terminal. The terminal strip comprises of many terminals. The procedure to insert terminal strips is given next.

- Click on the **Insert terminal strip** tool from the **Insertion** panel in the **Cabinet layout** tab of the **Ribbon**. The **Terminal strip selector** dialog box will be displayed; refer to Figure-9.

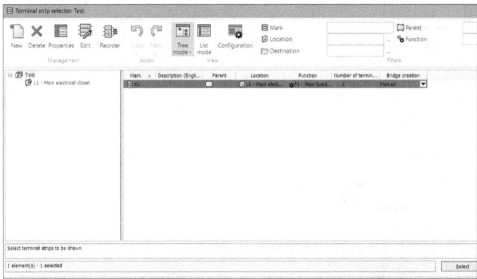

Figure-9. Terminal strip selector dialog box

Creating New Terminal Strip

- Click on the **New** tool from the **Management** panel in the dialog box. The **Component properties** dialog box will be displayed; refer to Figure-10.
- Specify desired description & properties for the terminal strip and click on the **OK** button from the dialog box. The terminal strip will get added in the **Terminal strip selector** dialog box.

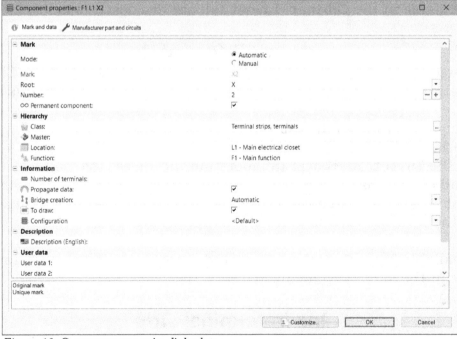

Figure-10. Component properties dialog box

- Select the newly created strip from the dialog box and click on the **Edit** button. The **Terminal strip editor** dialog box will be displayed as shown in Figure-11.

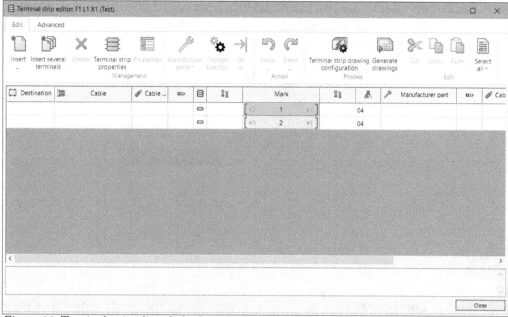

Figure-11. Terminal strip editor dialog box

• Click on the **Insert** tool from the **Management** panel in the **Edit** tab of the dialog box. A list of tools will be displayed; refer to Figure-12.

• Select the **Insert accessory** or **Insert accessory manufacturer part** option to add an accessory in terminal strip. Click on the Insert terminal option to add a new terminal in strip.

• To add multiple strip, click on the **Insert several terminals** tool from the **Management** panel. The **Multiple insertion** dialog box will be displayed as shown in Figure-13.

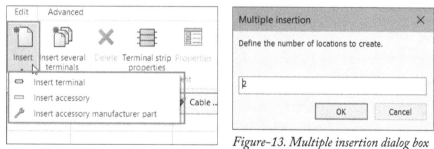

Figure-12. Insert drop-down

Figure-13. Multiple insertion dialog box

• Specify desired number of terminals to be added in the strip and click on the **OK** button. Like in Figure-14, we have created 6 terminals in a terminal strip.

Destination		Cable	Cable ...				Mark			
							1			
							2			
							3			
							4			
							5			
							6			

Figure-14. Terminal strip created

• Select desired terminal from the **Mark** column and click on the **Assign manufacturer parts** button from the **Manufacturer parts** drop-down; refer to Figure-15.

The **Manufacturer part selection** dialog box will be displayed. Select desired manufacturer part for each terminal.

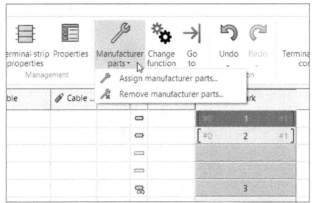

Figure-15. Assign parts tool

- To create a drawing file of terminal strip with specified configuration, click on the **Terminal strip drawing configuration** tool from the **Process** panel. The **Terminal strip drawing configuration** dialog box will be displayed; refer to Figure-16.

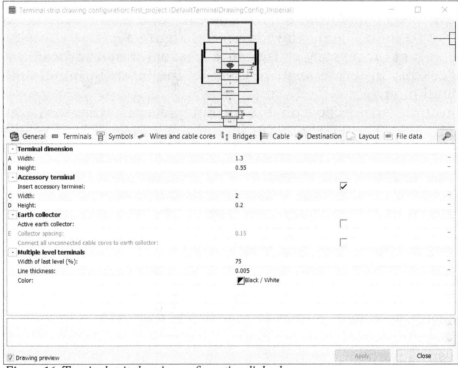

Figure-16. Terminal strip drawing configuration dialog box

- Set the parameters for the terminal strip and click on the **Apply** button.
- Click on the **Close** button to exit the dialog box.
- If you want to create a separate drawing for the terminals then click on **Generate drawings** tool. The **Selection of** dialog box will be displayed; refer to Figure-17.
- Select desired book & folder and click on the **OK** button from the dialog box. The drawing will be added in the selected book and summary will be displayed.
- Close the summary box and click on the **Close** button from the **Terminal strip editor** dialog box.

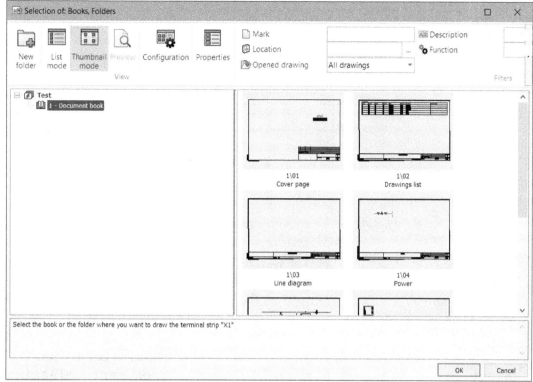

Figure-17. Selection of dialog box

- Now, click on the **Select** button from the **Terminal strip selector** dialog box. A terminal will get attached to the cursor; refer to Figure-18.

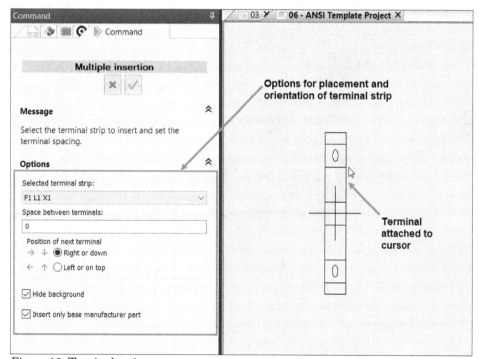

Figure-18. Terminal setting

- Set desired parameters for successive terminals in the **Options** rollout in **Command Browser** displayed.
- Click to specify the first terminal, rest of the terminals will be created automatically; refer to Figure-19.

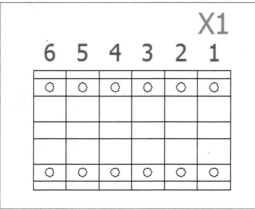

Figure-19. Terminals created

Add cabinet

Cabinet is an enclosure used to pack all the panel components so that the components inside cabinet are safe from external factors like dirt, water, and so on. The procedure to add a cabinet is given next.

- Click on the **Add cabinet** tool from the **Add a new manufacturer part** panel in the **Cabinet layout** tab of the **Ribbon**. The **Manufacturer part selection** dialog box will be displayed as discussed earlier.
- Set desired filters in the dialog box and click on the **Search** button to display the available cabinets; refer to Figure-20.

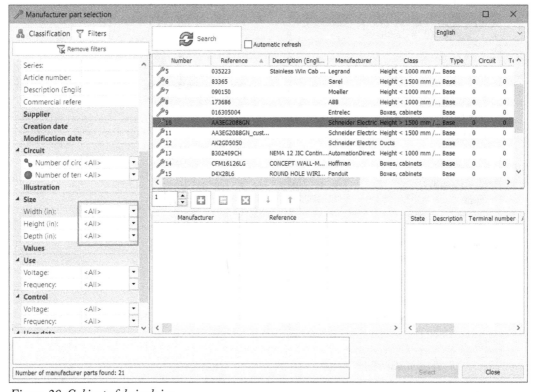

Figure-20. Cabinet of desired size

- Double-click on the cabinet to be used and then click on the **Select** button. The cabinet will get attached to the cursor.

- Click to place the cabinet at desired location. Note that if you have already placed components of panel layout and later place cabinet on them then it might hide the components in background. Here, we need to change the order of cabinet and set it to back.
- Select the newly placed cabinet enclosure and click on the **Set background** tool from the **Order** panel in the **Cabinet layout** tab of **Ribbon**; refer to Figure-21. The enclosure will move to the back.

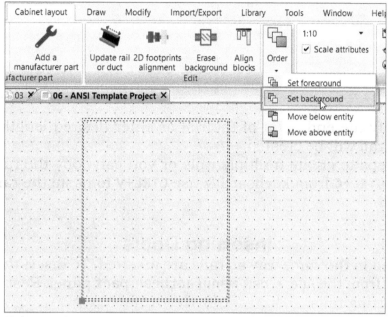

Figure-21. Set background tool

Although we have discussed this tool later in this chapter, but this is the first step when we start working on cabinet layout. So, while working in your organization, you should first place the cabinet, then rails and then the other components. Now, we will discuss about inserting rails in the cabinet.

Inserting rail

Rail is a strip of metal which holds various components like circuit breakers, contacts, and so on inside the cabinet. The procedure to insert rails in cabinet is given next.

- Click on the **Add rail** tool from the **Add a new manufacturer part** panel in the **Ribbon**. The **Manufacturer part selection** dialog box will be displayed.
- Set desired filter and click on the **Search** button. The list of rails available in the database will be displayed.
- Double-click on desired rail and then click on the **Select** button from the dialog box. The rail strip will get attached to the cursor.
- Specify the starting point of the rail by clicking inside the cabinet. You are asked to specify the length of the rail; refer to Figure-22.

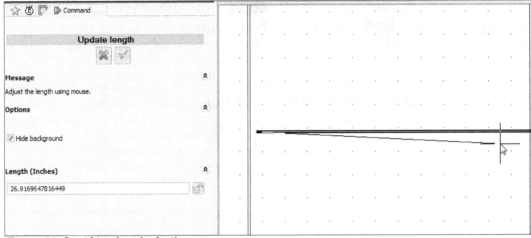

Figure-22. Specifying length of rail

- Click to specify the end point of the rail or enter desired length in the **Command Browser**. The rail will be created.
- Repeat the steps to create multiple rails or you can copy the rail by selecting it and pressing **CTRL+C** from keyboard & use **CTRL+V** to paste the rail. Click to place the copied rail.

Inserting Ducts

Ducts are inserted in the same way as rails are inserted. To insert the ducts, click on the **Add duct** tool from the **Add a new manufacturer part** panel. Rest of the procedure is similar to inserting rails.

Aligning Footprints

Once you have created footprints of components in the cabinet, the next step is to align them properly. In SolidWorks Electrical, there is a separate tool to perform this task. The procedure to align footprints is given next.

- Select the footprints that you want to align either by window selection or by holding the **CTRL** key while selecting footprints.
- Click on the **2D footprints alignment** tool from the **Edit** panel in the **Cabinet layout** tab of the **Ribbon**. The **Align 2D footprints** page will be displayed in the **Command Browser**; refer to Figure-23.

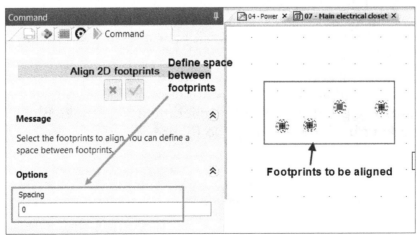

Figure-23. Aligning footprints

- Specify the spacing between consecutive footprints in the edit box available in the **Command Browser** and click on the **OK** button. The footprints will get aligned by specified distance; refer to Figure-24.

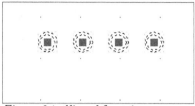
Figure-24. Aligned footprints

Note that once you are done with the designing of cabinet, its important to apply the dimensions so that it can be manufactured.

Erase background

The **Erase background** tool is used to remove background of selected footprint.

Align blocks

The **Align blocks** tool is used to align the selected footprints as per the selected reference block. The procedure to use this tool is given next.

- Click on the **Align blocks** tool from the **Edit** panel in the **Cabinet layout** tab of **Ribbon**. The **Align blocks Command Browser** will be displayed; refer to Figure-25.

Figure-25. Align blocks Command Browser

- Select the block whose insertion point is to be used as reference for alignment; refer to Figure-26 and click on the **OK** button from the **Command Browser**. The updated **Command Browser** will be displayed and you will be asked to select blocks for alignment.

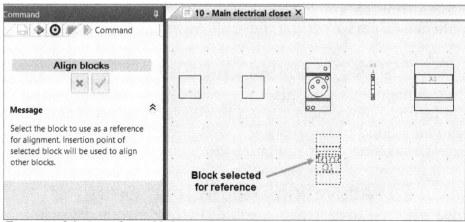
Figure-26. Selecting reference block

- Select desired option from the **Options** rollout of the **Command Browser** like Horizontal, Vertical, or Nearest button and select desired objects to be aligned while holding the **CTRL** key; refer to Figure-27.

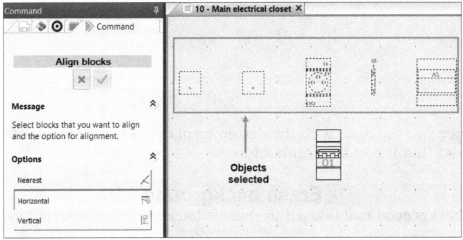

Figure-27. Objects selected for alignment

- Click on the **OK** button from the **Command Browser**. The selected objects will be aligned.

Defining Order of Parts

The options in the **Order** drop-down are used to define placement order of parts when they are placed one over the other; refer to Figure-28.

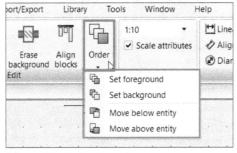

Figure-28. Order drop-down

After selecting the object, click on desired option from the **Order** drop-down.

Attribute Scale

The drop-down in **Scale** panel of **Cabinet layout** tab in **Ribbon** is used to set desired scale value for attributes and footprint. Make sure you have selected the **Scale attributes** check box before setting the scale so that text attributes are also scaled.

DRAWING OPTIONS

While creating dimensions and placing objects, you will find that cursor snap automatically to grid points, object snap points, and so on. The options to define snap settings and related parameters are available in the **Drawing options** panel of the **View** tab in the **Ribbon**; refer to Figure-29.

Activating/Deactivating Grid

The **Grid** toggle button is used to activate grids in the drawing. By default, this button is active in the drawing. Deselect the **Grid** toggle button to hide the grid.

Figure-29. Drawing options panel

Activating Orthomode

When orthomode is active then cursor will move only in horizontal or vertical direction while creating objects. Select the **Orthomode** toggle button to activate ortho mode.

Activating Snap Mode

The **Snap** toggle button is used to activate snapping of cursor to grid points. Select the **Snap** button to activate snapping of grid points.

Activating Line Thickness

The **Line thickness** toggle button is used to display thickness of lines and objects in schematic drawing.

Activating Object Snap

The **Object snap** toggle button is used to activate snapping of cursor to major points of object like center, end point, mid points, and so on. Select this button to activate snapping.

Setting Parameters for Object Snap

The **Parameters** button is used to define points to which cursor can snap when Object snapping mode is active. On clicking this button, the **Drawing parameters** dialog box is displayed; refer to Figure-30.

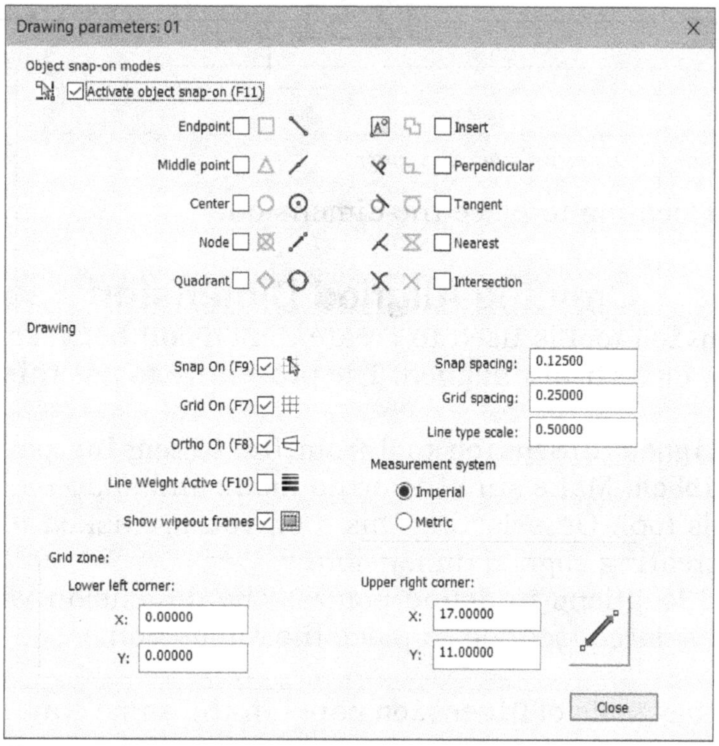

Figure-30. Drawing parameters dialog box

Select check boxes for key points to which you want the cursor to snap while create objects. You can also specify spacing for grids and snap points in the **Grid spacing** and **Snap spacing** edit boxes, respectively. Set desired value in **Line type scale** edit box if you want to scale up/down the line thickness by specified ratio in the drawing. Select the **Show wipeout frame** check box to display boundary rectangle around non-rectangular parts. From the **Grid zone** area of the dialog box, specify desired values for defining boundary within which grid points will be created. After setting desired parameters, click on the **Close** button from the dialog box.

CREATING DIMENSIONS

The tools in the **Dimension** panel are used to create various dimensions like linear dimension, aligned dimension, radius dimension, and so on. The dimensions are useful for fabrication of panel physically. Various tools in this panel are discussed next.

Creating Linear Dimension

The **Linear dimension** tool in **Dimension** panel is used to create linear horizontal or vertical dimension in the model. The procedure to use this tool is given next.

*   Click on the **Linear dimension** tool from the **Dimension** panel in the **Cabinet layout** tab of the **Ribbon**. The **Linear dimension** page will be displayed in the **Command Browser**.
*   Click at desired location in the drawing to specify start location of dimension. The other end of dimension will get attached to cursor.
*   Click to specify the end point of dimension line. The linear dimension will get attached to cursor; refer to Figure-31.

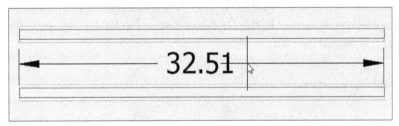

Figure-31. Dimension attached to cursor

*   Click at desired location to place the dimension.

Creating Aligned Dimension

The **Aligned dimension** tool is used to create dimension between two points which are not horizontally or vertically aligned. The procedure to use this tool is given next.

*   Click on the **Aligned dimension** tool from the **Dimension** panel in the **Cabinet layout** tab of **Ribbon**. Make sure the ortho mode and snap mode are deactivated before using this tool. On selecting this tool, you are asked to specify start and end points for creating aligned dimension.
*   Click at desired locations to define points. The dimension will get attached to cursor. Click at desired location to place the dimension.

You can use the other tools of **Dimension** panel in the same way.

CREATING TEXT LEADER

The **Text Leader** tool is used to mark text to an object with leader. The procedure to use this tool is given next.

- Click on the **Text Leader** tool from the **Annotation** panel in the **Cabinet layout** tab of the **Ribbon**. You will be asked to specify start point of the leader.
- Click at desired location to define the start point. You will be asked to specify end point of leader arrow.
- Click to specify the end point of arrow. You will be asked to specify end point of leader landing.
- Click at desired location to specify the end point of landing. The **Create a multiline text** dialog box will be displayed; refer to Figure-32.

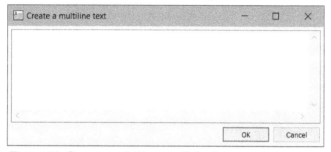

Figure-32. Create a multiline text dialog box

- Specify desired text in the dialog box and click on the **OK** button. The text will be created with leader; refer to Figure-33.

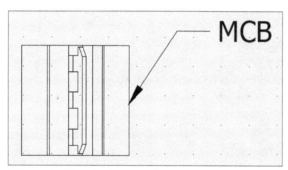

Figure-33. Text leader created

You can use the **Block leader** tool in the same way as **Text Leader** works. The block leader is used to mark view number and tag number of an object in the drawing.

Till this point, we have learnt almost all the tools necessary for the use of SolidWorks Electrical in industry. In the next chapter, we will perform some exercises using SolidWorks Electrical.

FOR STUDENT NOTES

Chapter 7

Practical and Practice

Topics Covered

The major topics covered in this chapter are:

- *Practice questions and practical*

INTRODUCTION

In the previous chapters, we have discussed various electrical drawing creation tools but it is easy to follow the procedure given in the book and perform the do this and do that format. Does it really help us in our general work in industry? My answer would be No. In this chapter, we will practically apply the tools and techniques to understand the application of SolidWorks Electrical in the industry.

PRACTICAL 1

Starting with the most basic control circuit, we will create a 3-Wire Control circuit for low voltage protection as given in Figure-1.

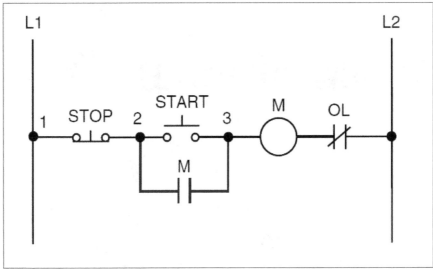

Figure-1. Practical 1

Steps:

Starting a new Project

- I hope you have started SolidWorks Electrical now!! By default, the **Electrical Project Management** dialog box will be displayed on starting the software. If it is not displayed in your case then click on the **Electrical Project** tool from the **Management** panel in the **Home** tab of the **Ribbon** to display the **Electrical Project Management** dialog box; refer to Figure-2.
- Click on the **New** button from the **Management** panel in the **Electrical Project Management** dialog box. You are asked to select desired template from the **Create a new project** dialog box.
- Select the **ANSI** template and click on the **OK** button from the dialog box. The **Project language** dialog box will be displayed.
- Select **English** as language and click on the **OK** button from the dialog box. The **Electrical Project** dialog box will be displayed.
- Specify the user data as per the instructions given by your trainer.
- Once you click on the **OK** button after specifying the data, the newly created project gets added in the project list of **Electrical Project Management** dialog box.
- Make sure that the newly created project is open and then close the dialog box by selecting the **Close** button.

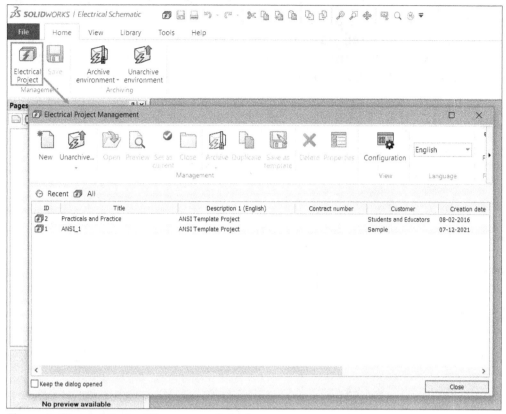

Figure-2. Electrical Project Management dialog box

Creating Control Circuit Wires

- Double-click on the **05-Control** drawing from the **Pages Browser**. The drawing will open; refer to Figure-3.

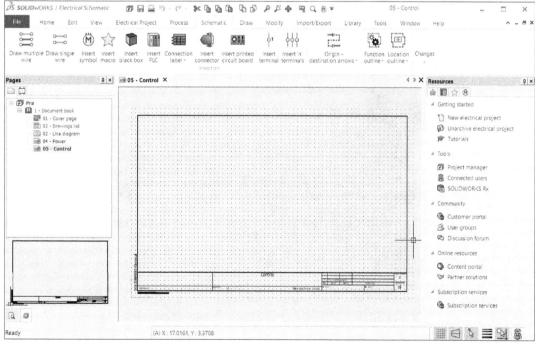

Figure-3. Control drawing

- Click on the **Draw single wire** tool from the **Insertion** panel in the **Schematic** tab of the **Ribbon**. The **Electrical wires** page will be displayed in the **Command Browser**.

- Enter the number of lines as **2** in the **Number of lines** spinner cum edit box in the **Command Browser**; refer to Figure-4.
- Click in the **Space between lines** edit box and specify the value as **8**.
- Click in the drawing area and create the wire lines as shown in Figure-5 and exit the tool by pressing the **ESC** button.

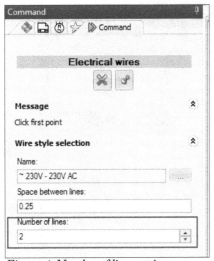

Figure-4. Number of lines setting

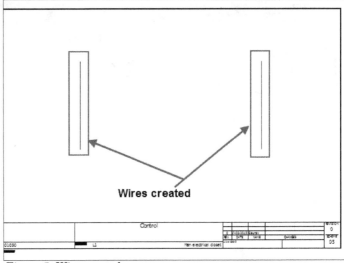

Figure-5. Wires created

- Click again on the **Draw single wire** tool and set the number of lines as **1**.
- Draw a wire connecting the earlier created wires at mid points as shown in Figure-6.

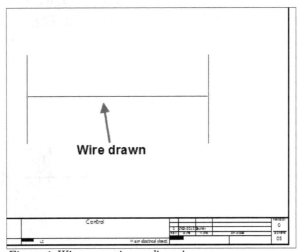

Figure-6. Wire connecting earlier wires

- Similarly, create a branch circuit in the connecting wire as shown in Figure-7.

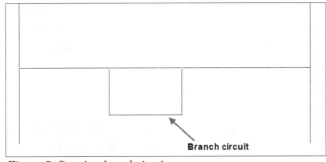

Figure-7. Creating branch circuit

Note that you need to clear **Grid** and **Snap** buttons from the **Drawing** options panel in the **View** tab of **Ribbon** because you need to click on wires at non grid points. Click on the **Parameters** button and make sure **Nearest** check box is selected for snap points in **Drawing parameters** dialog box. To activate object snapping, select the **Object snap** button.

Placing components

- Click on the **Insert symbol** tool from the **Insertion** panel in the **Schematic** tab of the **Ribbon**. The **Symbol insertion** page will be displayed in the **Command Browser**.
- Click on the **Other symbol** button from the page. The **Symbol selector** dialog box will be displayed; refer to Figure-8.

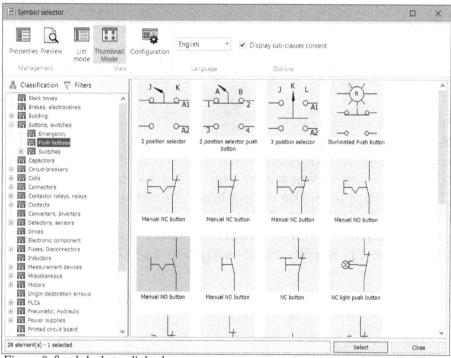

Figure-8. Symbol selector dialog box

- Select the **NC push button-Single Circuit - Momentary Contact** symbol from the **Buttons, switches > Push buttons** category in the dialog box and click on the **Select** button. The symbol will get attached to the cursor.
- Click on the wire to place the symbol at location as shown in Figure-9. The **Symbol properties** dialog box will be displayed as shown in Figure-10.

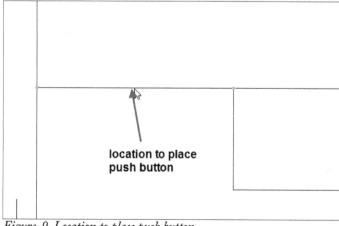

location to place
push button

Figure-9. Location to place push button

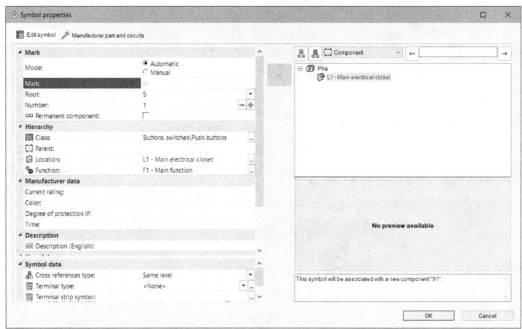

Figure-10. Symbol properties dialog box

- Click in the **Description (English)** edit box and enter **STOP**.
- Note that L1 is the location code for the symbol which means that while manufacturing, the component will be placed in L1 - Main electrical closet location.
- Click on the **OK** button from the dialog box. The symbol will be placed.
- Again, click on the **Insert symbol** button from the **Insertion** panel in the **Ribbon** and place the **NO push button - Single Circuit - Momentary Contact** symbol as shown in Figure-11. Specify the description as **START**.

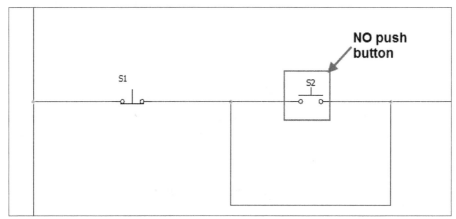

Figure-11. NO Push button placed

- Similarly, place the other symbols in the circuit; refer to Figure-12.

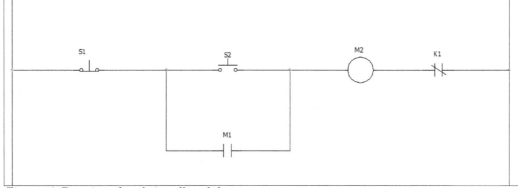

Figure-12. Drawing after placing all symbols

PRACTICE SET 1

On the basis of above practical, below are some drawings given in Figure-13, Figure-14, Figure-15, and Figure-16 that you can practice:

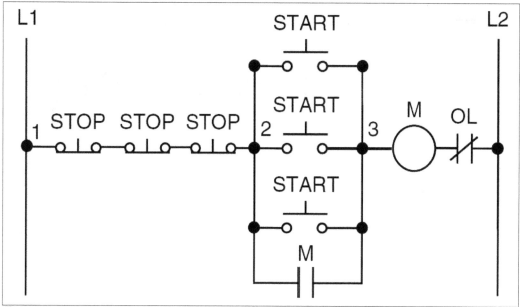

Figure-13. Practice 1

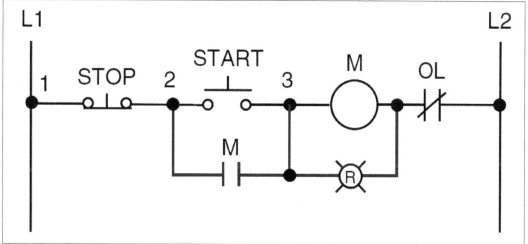

Figure-14. Practice 2

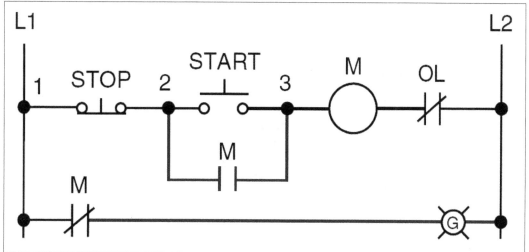

Figure-15. Practice 3

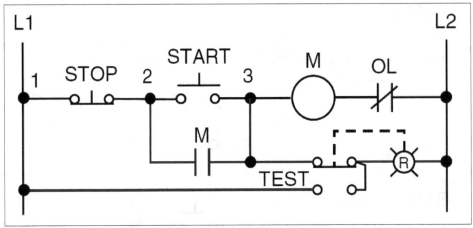

Figure-16. Practice 4

PRACTICAL 2

Create a transformer circuit with fuses as shown in Figure-17.

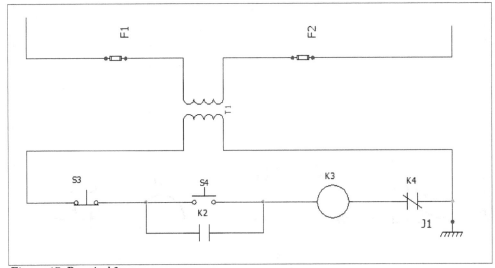

Figure-17. Practical 2

Starting New Drawing

I hope you are working in the same project that we have started in the Practical 1. Now, we will add another drawing the project.

•   Open the project created in Practical 1. Click on the down arrow below **New** button in the **Electrical Project** panel in the **Electrical Project** tab of the **Ribbon**. A list of tools will display; refer to Figure-18.

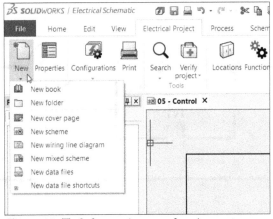

Figure-18. Tools for creating new drawings

- Click on the **New scheme** tool from the list. A new drawing will be added in the **Pages browser** and opened it.
- Right-click on the new drawing from **Pages browser** and select Properties option. The **Drawing** dialog box will be displayed.
- Specify the description as **Transformer Fused Circuit** in the **Description (English)** field of the dialog box; refer to Figure-19. Click on the **OK** button from the dialog box. The drawing will be modified in **Pages browser**; refer to Figure-20.

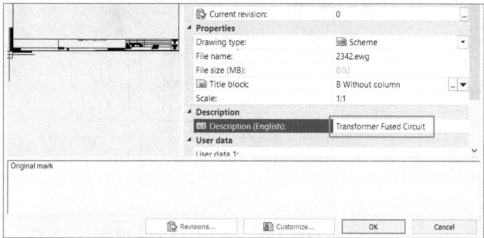
Figure-19. Specifying description

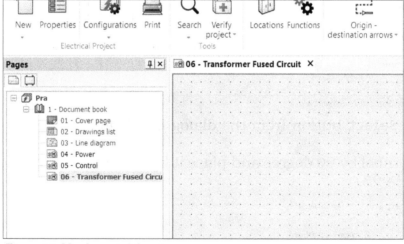
Figure-20. Newly created drawing

Inserting Wires

- Click on the **Draw single wire** tool from the **Insertion** panel in the **Schematic** tab of the **Ribbon** and draw a 230V AC wire as shown in Figure-21.

Figure-21. Wire to be drawn

Inserting transformer

- Click on the **Insert symbol** tool from the **Insertion** panel in the **Schematic** tab of the **Ribbon**. The **Symbol insertion** page will be displayed in the **Command Browser**.
- Click on the **Other symbol** button from the page and select the single phase transformer as shown in Figure-22.

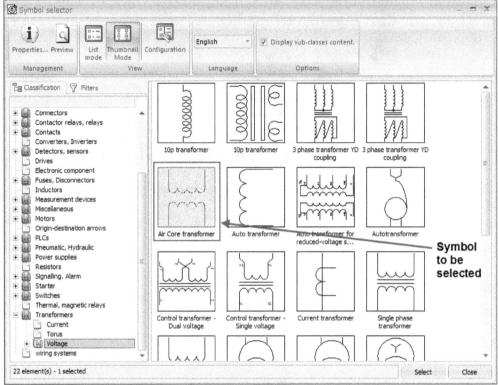

Figure-22. Symbol to be selected

- Click on the **Select** button from the dialog box. The symbol will get attached to the cursor.
- Rotate the symbol by 90 degree and place it by selecting the end point of the wire; refer to Figure-23.

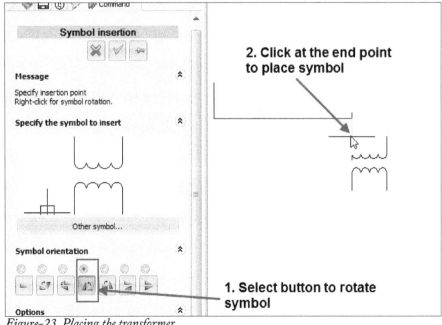

Figure-23. Placing the transformer

Drawing Rest of the Wires

• Click again on the **Draw single wire** tool from the **Schematic** tab in **Ribbon**. The options related to wiring will display in the **Command Browser**.

• Make sure that number of lines is set as **1** and **Wire style** as **230V AC** in the **Command Browser**. Click at the open end point of the transformer and draw the wiring as shown in Figure-24.

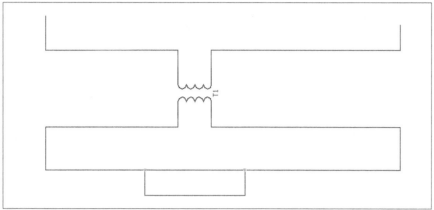

Figure-24. Wiring drawn

Inserting other symbols

• Click again on the **Insert symbol** tool from the **Schematic** tab in the **Ribbon** and one by one insert the symbols as shown in Figure-25.

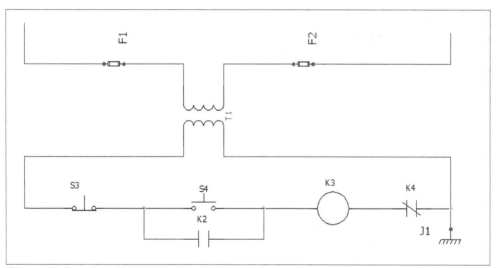

Figure-25. Inserting all components

Updating project data

Once we have done all the insertions, the last step is to update data of project which includes wire numbering, cable references, and so on.

• Click on the **Update data** tool from the **Management** panel in the **Process** tab of the **Ribbon**. The **Wizard to update project data** dialog box will be displayed; refer to Figure-26.

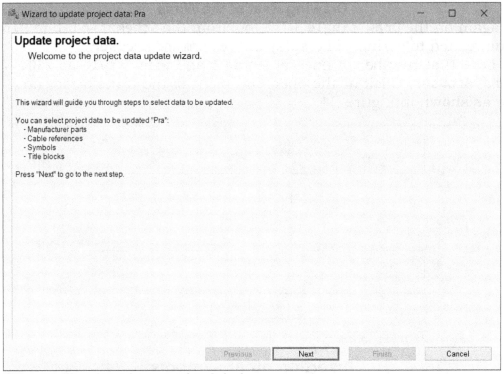

Figure-26. Wizard to update project data dialog box

- Set desired parameters and click on the **Next** buttons in dialog box. Click on the **Finish** button from the dialog box to update all the parameters.
- To display various markings like wire number; click on the down arrow next to **Show texts** button from the **Changes** panel in the **Schematic** tab of the **Ribbon**. Two tools will be displayed as shown in Figure-27.

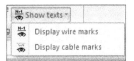

Figure-27. Tools to display marks

- Click on the **Display wire marks** tool from the list. You are asked to select the wires for which you want to display the marks. Select all the wires by using the window selection and press **ENTER**. The wire numbering will be applied to the wires; refer to Figure-28.

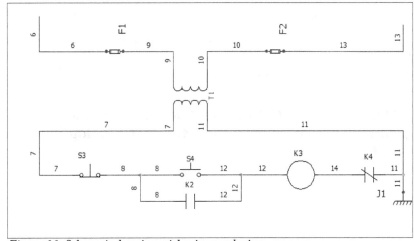

Figure-28. Schematic drawing with wire numbering

PRACTICE 5

On the basis of above practical, create the schematics of drawings shown in Figure-29 and Figure-30.

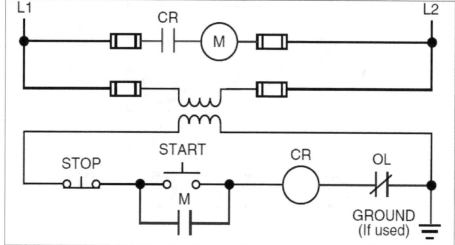

Figure-29. Practice 5

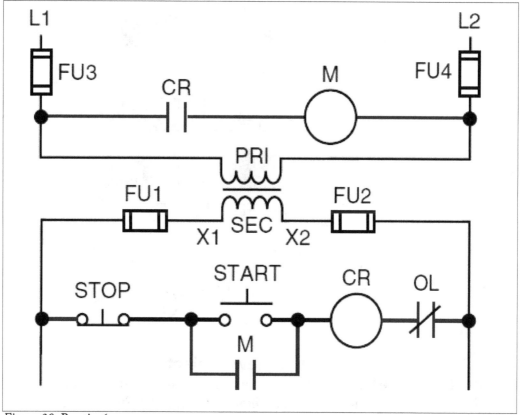

Figure-30. Practice 6

PRACTICE 6

Create panel drawing and bill of material for all the practical and practice questions discussed so far.

PRACTICAL 3

Create schematic for line diagram given in Figure-31.

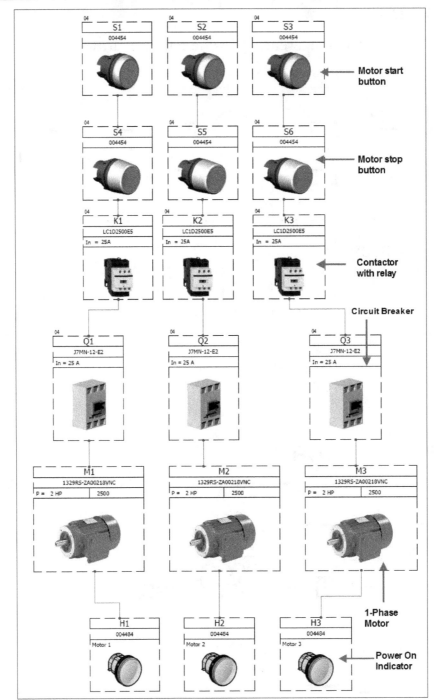

Figure-31. Practical 3

Starting a new Project

- Click on the **Electrical Project** tool from the **Management** panel in the **Home** tab of the **Ribbon** to display the **Projects Manager**; refer to Figure-32.

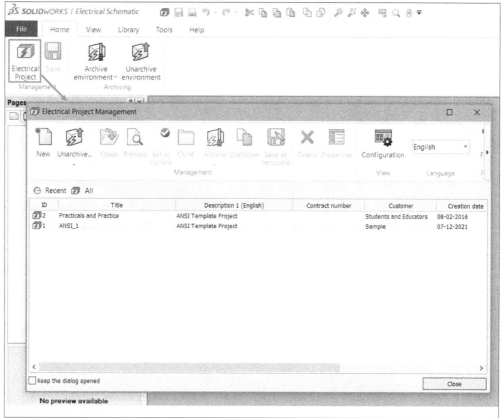

Figure-32. Electrical Project Management dialog box

- Click on the **New** button from the **Management** panel in the **Electrical Project Management** dialog box. You are asked to select desired template from the **Create a new project** dialog box.
- Select the ANSI template and click on the **OK** button from the dialog box. The **Project language** dialog box will be displayed.
- Select **English** as language and click on the **OK** button from the dialog box. The **Electrical Project** dialog box will be displayed.
- Specify the user data as per the instructions given by your tutor.
- Click on the **OK** button after specifying the data, the newly created project gets added in the project list of **Electrical Project Management** dialog box.
- Make sure that the newly created project is open and then close the **Projects Manager** by selecting the **Close** button.

Creating Line Diagram

- Double-click on the Line diagram drawing from the **Pages Browser** in the left of the application window. The drawing will open if not opened earlier.
- Right-click on the file name from the **Documents Browser** and select the **Replace** option in the **Title Block** cascading menu in the shortcut menu displayed; refer to Figure-33. The **Title block selector** dialog box will be displayed; refer to Figure-34.

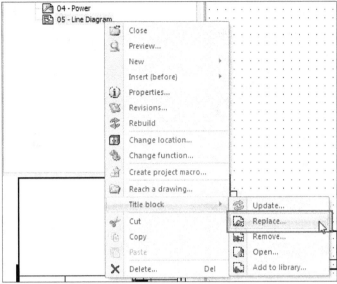

Figure-33. Replace option for title block

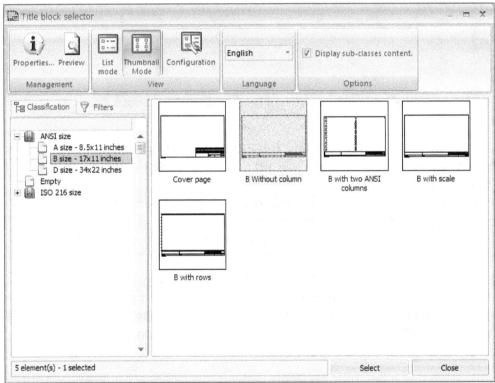

Figure-34. Title block selector dialog box

- We need a bigger size worksheet to create our line diagram, so click on the **D size - 34x22 inches** option from the left area of the dialog box. The title blocks with D size will be displayed on the right in the dialog box.
- Double-click on the **D without column without row option** in the dialog box. The template will change accordingly.

Adding Components in Line Diagram

- Click on the **Insert symbol** button from the **Insertion** panel in the **Line diagram** tab of **Ribbon**. The **Symbol insertion Command Browser** will be displayed; refer to Figure-35.
- By default, the symbol earlier used is displayed in the **Command Browser**. Click on the **Other symbol** button to display the **Symbol selector** dialog box.

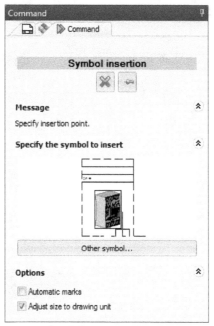

Figure-35. Symbol insertion CommandManager

- Click on the **Buttons, switches** category in the **Classification** tab at the left in the dialog box. The symbols related to buttons and switches will be displayed; refer to Figure-36.

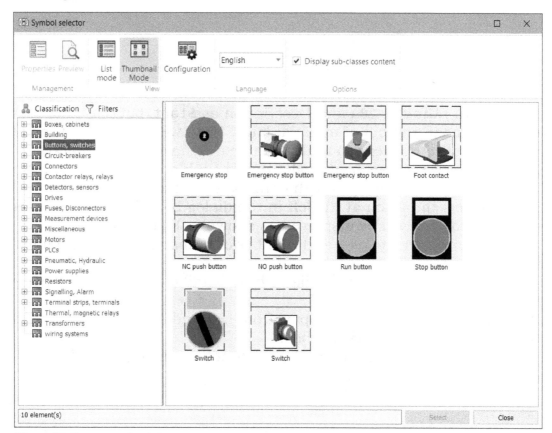

Figure-36. Buttons and switches

- Double-click on **NO push button** symbol from the dialog box. The symbol gets attached to cursor.
- Place the symbol near the top-left corner so that other symbols can be placed at the right and below this symbol.

- On placing the symbol, **Symbol properties** dialog box is displayed; refer to Figure-37.

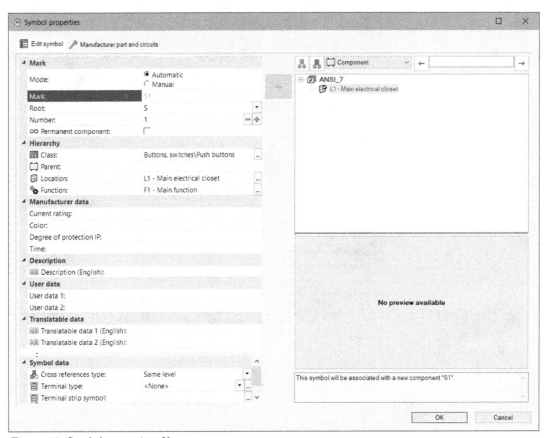

Figure-37. Symbol properties of button

- Set the current rating as **20A** and Description as **Start button**. Note that these values will come in the bill of material later. Now, we will specify the details of manufacturer of component so that it can be purchased in the market.
- Click on the **Manufacturer part and circuits** tab at the top in the dialog box. The dialog box will be displayed with options related to manufacturer.
- Click on the **Search** button at the left in the dialog box. The **Manufacturer part selection** dialog box will be displayed as shown in Figure-38.
- Set the number of circuits as **1** and number of terminals as **2** in the filters area; refer to Figure-39.
- Click on the **Search** button in the dialog box. Detail of manufacturer part will be displayed; refer to Figure-40.
- Double-click on the manufacturer part from Legrand manufacturer and click on the **Select** button from the dialog box.
- Click on the **OK** button from the **Symbol properties** dialog box. The symbol will be added in drawing for manufacturer part.

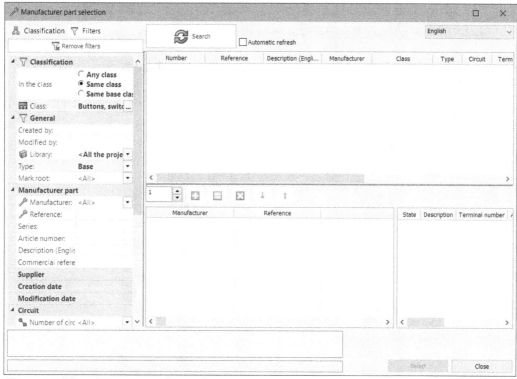

Figure-38. Manufacturer part selection dialog box

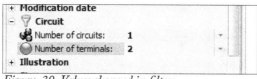

Figure-39. Values changed in filters

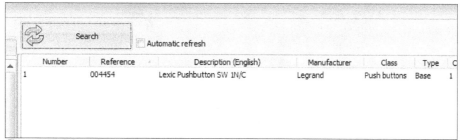

Figure-40. Manufacturer part detail

Similarly, insert the other components for first motor circuit with descriptions as given next.

NC Push button: **20A** (current rating), **Stop button** (Description), **Legrand** (Manufacturer), **004454** (Reference).

Contactor Relay: **25A** (current rating), **Relay 1** (Description), **Schneider Electric** (Manufacturer), **LC1D25500E5** (Reference).

Circuit Breaker: **25A** (current rating), **Circuit-breaker 1** (Description), **Omron** (Manufacturer), **J7MN-12-E2** (Reference).

Motor: **2HP** (Power), **2500** (Speed), **Motor** (Description), **Allen-Bradley** (Manufacturer), **1329RS-ZA00218VNC** (Reference).

Signalling, Alarm/Luminous: **5A** (Power), **Red** (Color), **220 V** (Voltage), **Motor ON** (Description), **Legrand** (Manufacturer), **004484** (Reference).

After putting all the parts, drawing should display as shown in Figure-41.

- Select all these symbols by using the cross-selection and press **CTRL** + **C** from the keyboard. All the symbols will be copied in system memory.
- Press **CTRL** + **V** from the keyboard and paste it two times as shown in Figure-42.

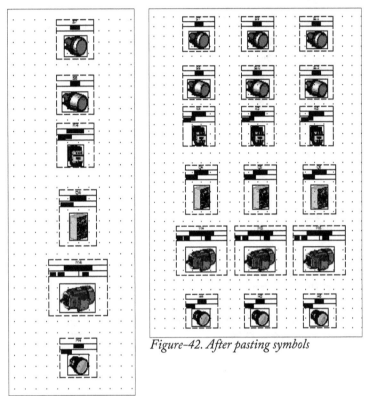

Figure-42. After pasting symbols

Figure-41. Symbols placed in drawing

Connecting Cable

- Click on the **Draw cable** tool from the **Insertion** panel in the **Line diagram** tab of **Ribbon**. The **Draw a cable Command Browser** will be displayed; refer to Figure-43.

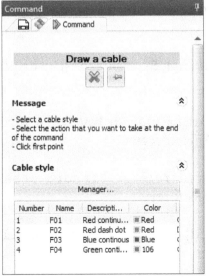

Figure-43. Draw a cable Command Browser

- Click at the bottom of S1 switch and then at the top of S4 switch; refer to Figure-44.

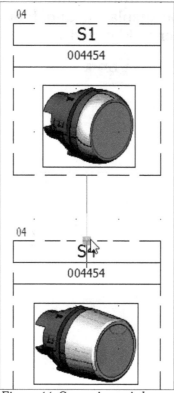

Figure-44. Connecting switches with wire

- Similarly, connect the other components in the line diagram; refer to Figure-45.

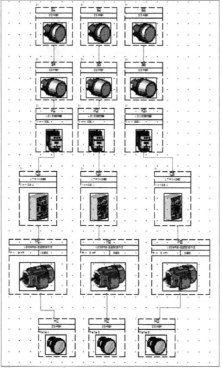

Figure-45. Connecting components in the line diagram

Based on this line diagram, we will now create schematic for the circuits.

Creating Schematic

- Double-click on the schematic drawing named **Power** from the **Pages Browser**. Blank drawing page will be displayed and tools related to schematic will be displayed in the **Ribbon**; refer to Figure-46.

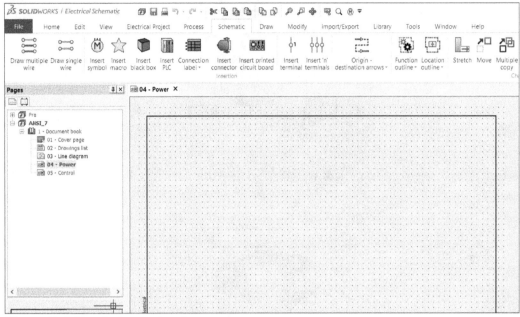

Figure-46. Schematic drawing interface

- Click on the **Draw single wire** tool from the **Insertion** panel in the **Schematic** tab of **Ribbon**. The **Electrical wires Command Browser** will be displayed; refer to Figure-47.
- Click on the **Browse** button next to **Name** field in the **Command Browser**. The **Wire style selector** dialog box will be displayed; refer to Figure-48.

Figure-47. Electrical wires Command Browser

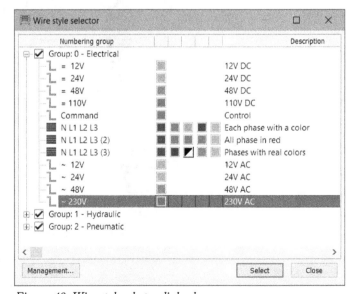

Figure-48. Wire style selector dialog box

- Click on the **Management** button at the bottom of **Wire style selector** dialog box. The **Wire style management** dialog box will be displayed; refer to Figure-49.

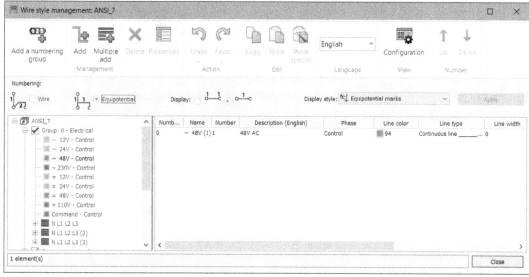

Figure-49. Wire style management dialog box

- Click on the **Add** button from the **Management** panel in the dialog box. A new wire will be added in the list; refer to Figure-50.

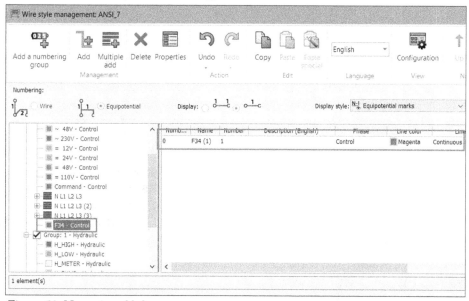

Figure-50. New wire added

- Right-click on the newly added wire and select the **Properties** option from the shortcut menu displayed; refer to Figure-51. The **Wire style** dialog box will be displayed; refer to Figure-52.

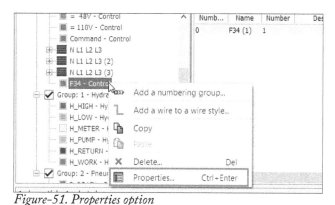

Figure-51. Properties option

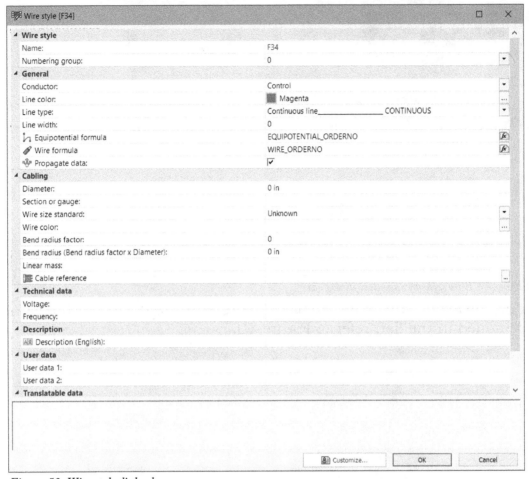

Figure-52. Wire style dialog box

- Click in the **Name** field and specify the name as **230V AC**. Similarly, set diameter as **2.5 mm**, Line color as **Red**, Wire size standard as **Section (mm²)**, Bend radius (x Diameter) as **25**, Voltage as **230 V** and Frequency as **60 Hz**.

- Click on the **OK** button from the dialog box. The wire will be created. Click on the **Close** button to exit the dialog box.

- Select the newly created wire from the **Wire style selector** dialog box and click on the **Select** button. The new wire will become active.

- Draw the wire as shown in Figure-53.

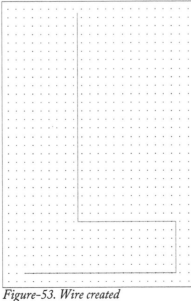

Figure-53. Wire created

Inserting Schematic Symbol

- Click on the **Insert symbol** tool from the **Insertion** panel in the **Schematic** tab of **Ribbon**. The **Symbol insertion Command Browser** will be displayed.
- Click on the **Other symbol** button. The **Symbol selector** dialog box will be displayed.
- Click on the **Buttons, switches** option from the left area of the dialog box and double-click on the **NO push button - Single Circuit - Momentary Contact** symbol; refer to Figure-54. The symbol gets attached to the cursor.

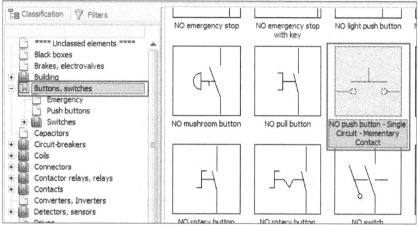

Figure-54. NO push button selected

- Click on the wire near the top end to place the symbol. The **Symbol properties** dialog box will be displayed; refer to Figure-55.

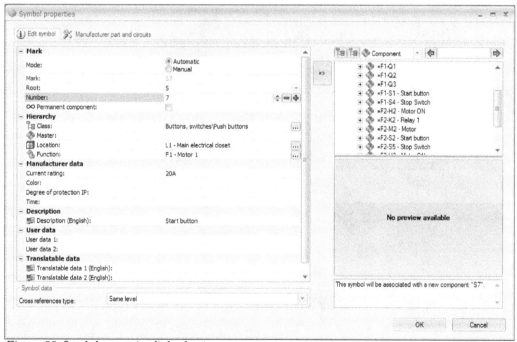

Figure-55. Symbol properties dialog box1

- Click on the **Same class** button above the **Component browser** in the dialog box; refer to Figure-56. The symbols of same class which have been used in the current project will be displayed.

- Click on the **S1-Start** button from the **Component Browser** and click on the **OK** button. The symbol will be added in the schematic drawing. Similarly, insert the NC push button matching its properties with S4 Stop switch; refer to Figure-57.

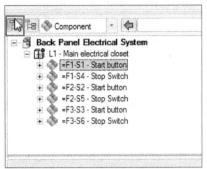

Figure-56. Same class button

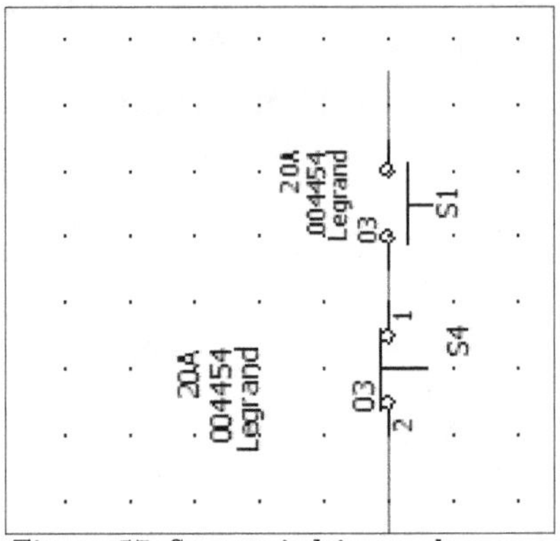

Figure-57. Stop-switch inserted

- Similarly, insert the other components for first motor circuit. Note that relay has two components in schematic relay contactor and relay coil. So, you need to insert both the components in the schematic; refer to Figure-58.
- Draw a neutral wire connecting to the N terminal of circuit breaker in circuit; refer to Figure-59.

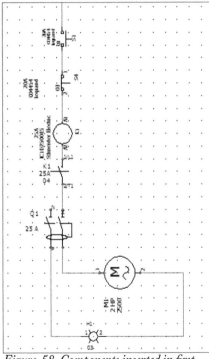

Figure-58. Components inserted in first motor circuit

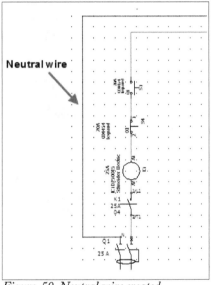

Figure-59. Neutral wire created

- Select this complete circuit by using window selection and make two copies of the circuit as shown in Figure-60.

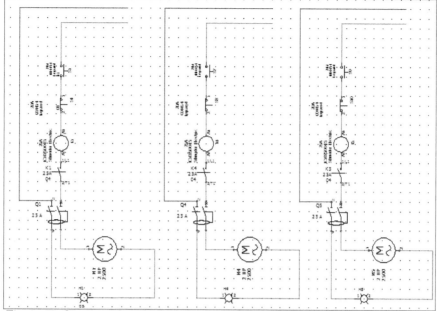

Figure-60. Circuit with multiple copies

Assigning Components to schematic symbols

- Right-click on the NO push button in second circuit and select the **Assign component** option from the shortcut menu; refer to Figure-61. The **Assign component CommandManager** will be displayed; refer to Figure-62.

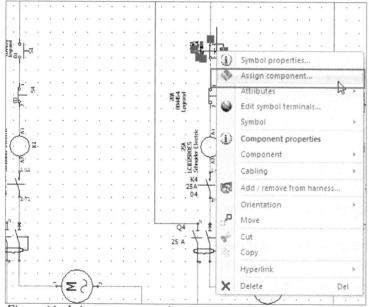

Figure-61. Assign component option

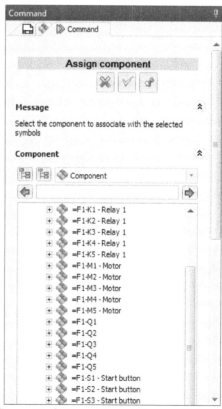

Figure-62. Assign component CommandManager

- Select the **S2 - Start button** component from the **CommandManager** and click **OK** button. The component properties of S2 will be assign to the component and S7 will be removed from database.
- Similarly, assign the component properties to the symbols in circuit as per the line diagram; refer to Figure-63.

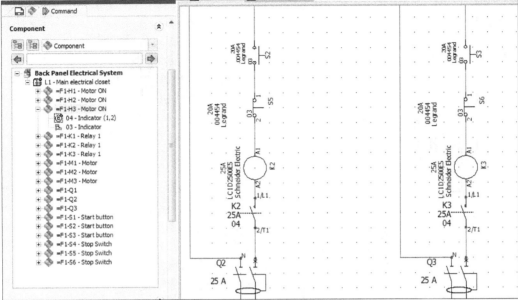

Figure-63. Assigning components to symbols as per line diagram

Connecting Terminal to wires

We are not given any power source in the question so we are going to connect all the open wires to terminals so that later power source can be directly connected to the terminals.

- Extend the wires using simple drag-drop functions on wires; refer to Figure-64.

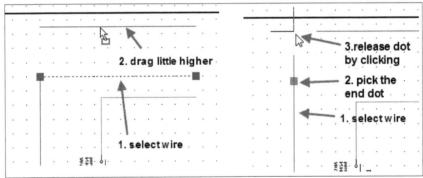

Figure-64. Drag-drop operations on wires

- After performing various drag-drop operations, make the wiring as shown in Figure-65.

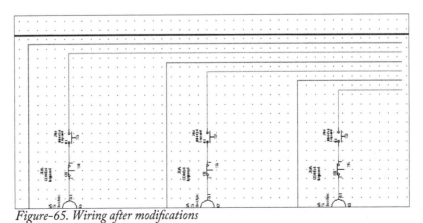

Figure-65. Wiring after modifications

- Click on the **Insert 'n' terminals** button from the **Insertion** panel in the **Schematic** tab of **Ribbon**. The **Terminal insertion Command Browser** will be displayed.
- Draw a vertical line intersecting the open ends of wires; refer to Figure-66. You are asked to define the orientation of terminals.

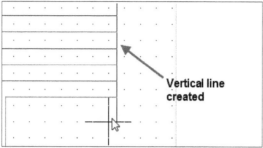

Figure-66. Vertical line created for terminals

- Click on the left of terminals. The **Terminal symbol properties** dialog box will be displayed; refer to Figure-67.

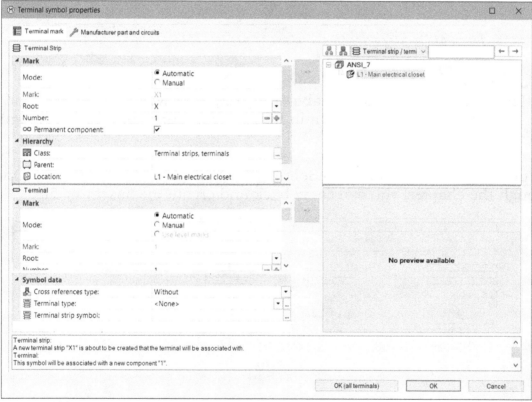

Figure-67. Terminal symbol properties dialog box

- Click on the **Manufacturer part and circuits** tab and select the **Entrelec** as manufacturer and Reference as **019532020** component from the list.
- Click on the **OK (all terminals)** button from the dialog box to create terminals. The schematic drawing will be displayed as shown in Figure-68.

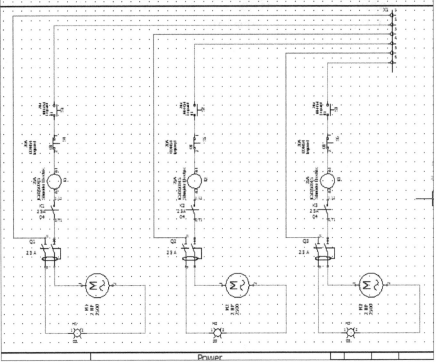

Figure-68. Drawing after adding terminals

PRACTICE 7

Create Line diagram, schematic, cabinet layout, SolidWorks Assembly, and bill of materials for motor circuit as shown in Figure-69, Figure-70, Figure-71, Figure-72, and Figure-73.

Note: You will learn about creating SolidWorks Assembly in next chapter. Once you go through the chapter, you can come back and create the assembly file.

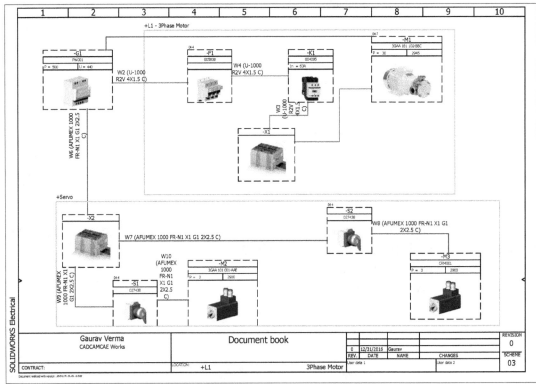

Figure-69. Line diagram

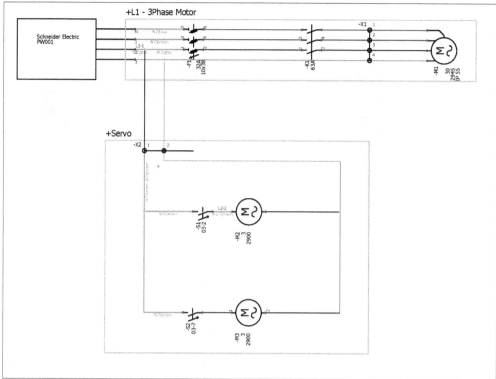

Figure-70. Schematic

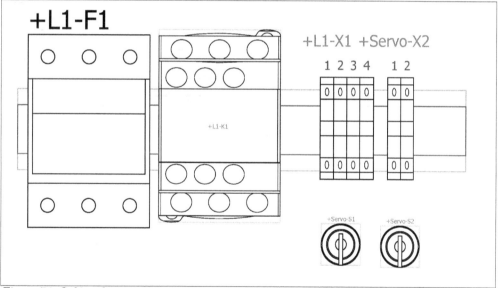

Figure-71. Cabinet layout

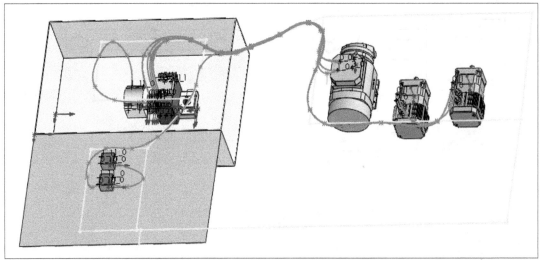

Figure-72. SolidWorks Assembly

ABB

Reference	Mark	Description	Quantity
3GAA 181 102-BBC	-M1	Motor 3-phases, type M3AA 180 LB ,2 poles - 400V YD - 50 Hz - High-output design, Flange-mounted	1

Crompton

Reference	Mark	Description	Quantity
CRM001	-M2 , -M3	Crompton 230V Single Phase Motor	2

Entrelec

Reference	Mark	Description	Quantity
010500220	-X1-1 , -X1-2 , -X1-3 , -X1-4 , -X2-1 , -X2-2	Simple terminal	6

Legrand

Reference	Mark	Description	Quantity
004095	-K1	Legrand Contactor 004095	1
005838	-F1	Mod TP 10x38mm Fuse Carrier	1
009213	-A2	Legrand Rail for Cabinet	1
035223	-A1	Stainless Win Cab 600X400X2	1
027438	-S1 , -S2	Cam switch - changeover switch with off 45° - PR 17 - 4P - 20 A - screw fixing	2

Schneider Electric

Reference	Mark	Description	Quantity
PW001	-G1	Schneider Electric Power Supply	1

Figure-73. Bill of Materials

Chapter 8

Electrical 3D

Topics Covered

The major topics covered in this chapter are:

- *SolidWorks Electrical 3D interface*
- *3D Electrical Parts*
- *Inserting Electrical Components*
- *CAD File Downloader*
- *Routing Wires*
- *Creating Routing Path*
- *Updating BOM Properties*

INTRODUCTION

SolidWorks Electrical Professional is combination of two interconnected electrical packages; SolidWorks 2D Electrical (Schematic) and SolidWorks 3D Electrical. Before this chapter, we have discussed about 2D part of SolidWorks Electrical. Now, we will discuss about 3D part of SolidWorks Electrical. Make sure that you have installed SolidWorks application and SolidWorks Electrical add-in for it. Before working with 3D electrical routing, we must have 3D parts that represent the symbols of SolidWorks 2D electrical.

CREATING SOLIDWORKS ELECTRICAL PART

Before creating any electrical part, you must have a model already created in SolidWorks. You can create a part or you can download it from 3dcontentcentral. com. Rest of steps are given next.

- Open the solid part that you want to make electrical part in SolidWorks; refer to Figure-1.

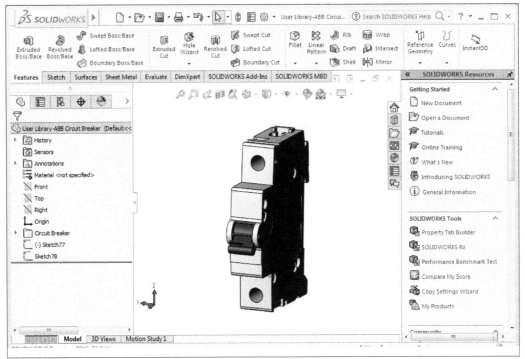

Figure-1. Part opened in SolidWorks

- Click on the **Add-Ins** option from the **Options** drop-down in the **Quick Access Toolbar**; refer to Figure-2. The **Add-Ins** selection box will be displayed; refer to Figure-3.

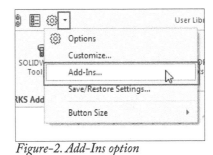

Figure-2. Add-Ins option

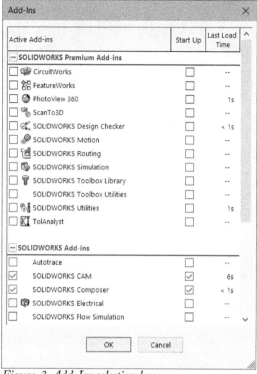

Figure–3. Add-Ins selection box

- Select the check box before **SOLIDWORKS Electrical** in the selection box and click on the **OK** button. The toolbar for SolidWorks Electrical 3D will be displayed; refer to Figure-4.

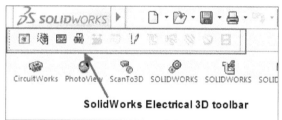

Figure–4. Solidworks electrical 3D toolbar

- Click on the **Electrical Component Wizard** tool from the **SolidWorks Electrical 3D** toolbar or click on the **Electrical Component Wizard** tool from the **Tools > SOLIDWORKS Electrical** menu. The **Routing Library Manager** dialog box will be displayed; refer to Figure-5.
- Select desired route type and component type from the **Routing Component Wizard** tab in the dialog box. For example, if you want to create a circuit breaker model which is attached to Din Rail in the cabinet then select the **Electrical** radio button from the **Route type** area and **Din Rail Component** radio button from the **Component type** area. Click on the **Next** button from the dialog box. The **Routing Functionality Points** page will be displayed; refer to Figure-6.

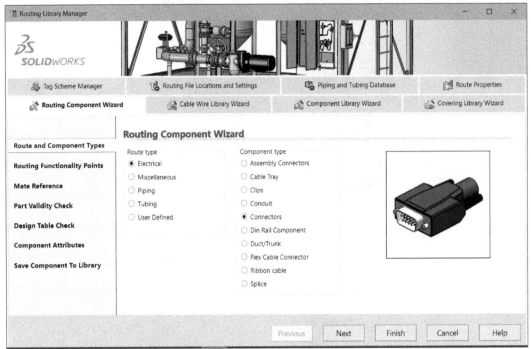

Figure-5. Routing Library Manager dialog box

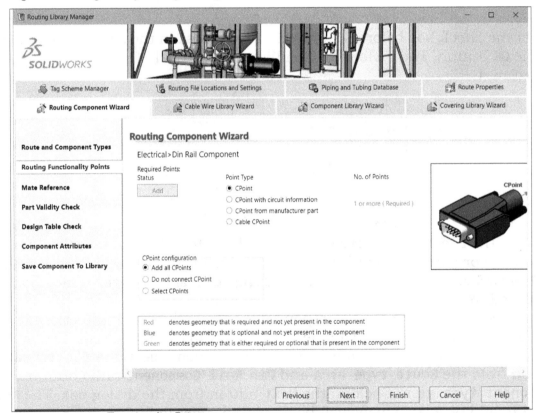

Figure-6. Routing Functionality Points page

- Select desired type of connection point to be created from the **Point Type** area. Select the **CPoint** radio button if you want to create connection point without any information. Select the **CPoint with Circuit information** radio button to define circuit along with connection point. Select the **CPoint from Manufacturer part** radio button if you want to specify connection points based on manufacturer part data. Select the **Cable CPoint** radio button is used to create connection point for cable. The most commonly used option is **CPoint from manufacturer part** radio button so, we will be using this option for demonstration.

- After selecting the **CPoint from Manufacturer part** radio button, click on the **Add** button. The **Create Connection Points PropertyManager** will be displayed; refer to Figure-7.

Figure-7. Create connection points Property Manager for connection references

- Click on the **Select manufacturer part** button from the **PropertyManager**. The **Manufacturer part selection** dialog box will be displayed; refer to Figure-8.

Figure-8. Manufacturer part selection dialog box

- Set desired filters, search, and select manufacturer part as required; refer to Figure-9.

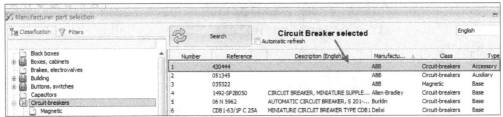
Figure-9. Manufacturer part selected

- On selecting the manufacturer part, terminals of the part are displayed in the **PropertyManager**; refer to Figure-10. Also, you are asked to select a point for the terminal.

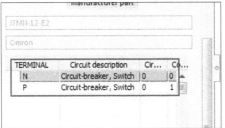
Figure-10. Terminals of part displayed in PropertyManager

- Right-click on the terminal and select the **Create Connection Point** option; refer to Figure-11. You are asked to select the point.

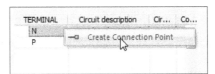
Figure-11. Create Connection Point option

- Click on the sketch point created on the part to make it terminal; refer to Figure-12.

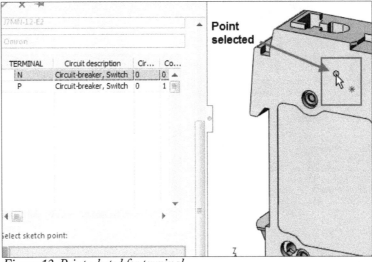
Figure-12. Point selected for terminal

- Similarly, create the other connection points and then click on the **OK** button from the **PropertyManager**. Make sure you have unpin the **PropertyManager** to exit the tool. The page in **Routing Library Manager** will be displayed with specified pins attached.
- Click on the **Next** button from the dialog box. The **Mate Reference** page will be displayed in the dialog box; refer to Figure-13.

- Select desired radio button from the **Reference Name** area. Select the **Default** radio button to select default placement reference set for the part as per the database. Select the **For Rail** radio button to define placement reference for rail in the cabinet. Select the **For Cabinet** radio button to place part directly on the cabinet face. Select the **For Cabinet Door** radio button to place the part on door of cabinet. Since we are creating Circuit Breaker part so we will use **For Rail** radio button as component will be placed on rail in the cabinet.

- After selecting **For Rail** radio button, click on the **Define Faces** button from the **Component Alignment Options** area of the dialog box. The **Define all the faces of the component PropertyManager** will be displayed; refer to Figure-14.

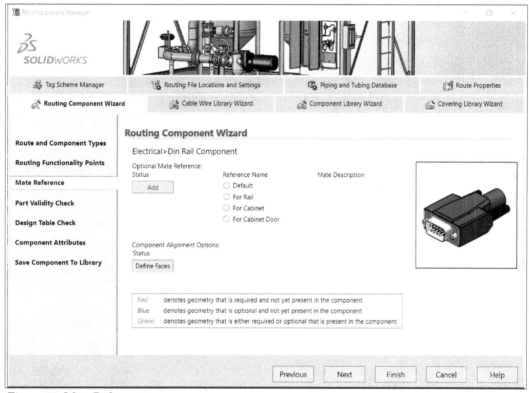

Figure-13. Mate Reference page

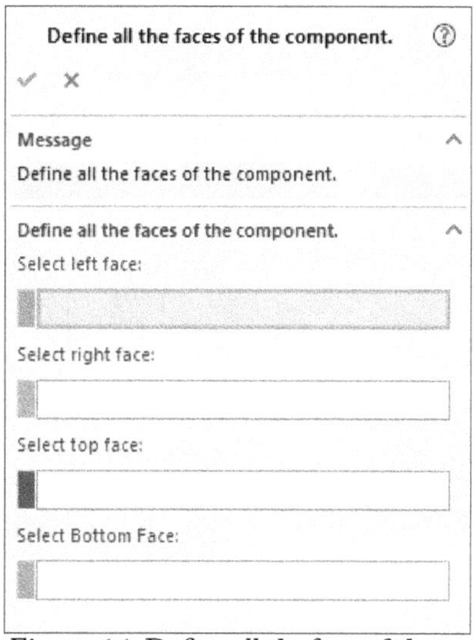

Figure-14. Define all the faces of the component PropertyManager

- One by one select the left, right, top, and bottom faces of the component; refer to Figure-15.

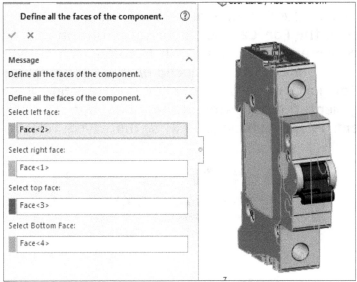

Figure-15. Faces selected from the model

- Click on the **OK** button from the **PropertyManager**. The **Electrical component wizard** will be displayed again.
- Click on the **Add** button from the **Mate Reference** page in the **Routing Library Manager** dialog box. The **Create mate reference PropertyManager** will be displayed; refer to Figure-16. Also, you are asked to select face for mating top face of the rail.
- Select the reference face for top face of the rail. You are asked to select reference face for front face of the rail.
- Select the reference face; refer to Figure-17.

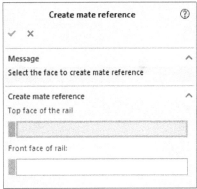

Figure-16. Create mate reference Prop-ertyManager

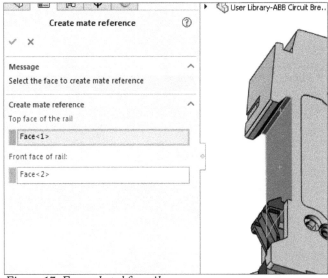

Figure-17. Faces selected for rail

- Click on the **OK** button from the **PropertyManager**. The **Electrical component wizard** dialog box will be displayed. The page will be displayed as shown in Figure-18.

Routing Component Wizard

Electrical > Din Rail Component

Optional Mate Reference:
Status

		Reference Name	Mate Description
Edit	Delete	TREWRAIL35-<1>	Primary Reference : present Secondary Reference : present
Add		○ Default	
		○ For Rail	
		○ For Cabinet	
		○ For Cabinet Door	

Route and Component Types

Routing Functionality Points

Mate Reference

Part Validity Check

Design Table Check

Component Attributes

Save Component To Library

Component Alignment Options:
Status

Define Faces

Figure-18. After specifying mate reference

- Click on the **Next** button from the dialog box. The **Part Validity Check** page of dialog box will be displayed; refer to Figure-19. If there is any information missing to create the part then it will be displayed under **Missing required items** area of the dialog box. If there is missing information, then specify the related parameter/ reference.

Figure-19. Part Validity Check page

- Click on the **Next** button from the dialog box. The **Design Table Check** page of the dialog box will be displayed; refer to Figure-20.

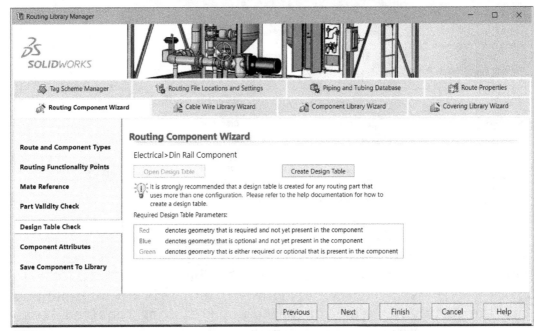

Figure-20. Design Table Check page

- Using a design table, you can create multiple instances of component with difference sized based on a design table. To create design table, click on the **Create Design Table** button. The **Excel Design Table PropertyManager** will be displayed; refer to Figure-21.

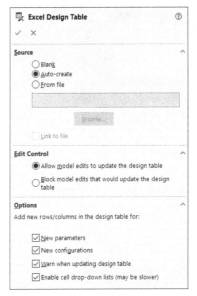

Figure-21. Excel Design Table PropertyManager

- Select the **Auto-create** radio button and set other parameters as desired. (You can learn more about Excel Design Table in our other book SolidWorks 2023 Black Book which is dedicated for Model design). After setting desired parameters, click on the **OK** button from the **PropertyManager**. The excel sheet of design table will be displayed and you will be asked to select dimensions for the table; refer to Figure-22.
- Select the dimensions to be used in the design table and click on the **OK** button. The dimensions will be displayed in the table.
- Set desired values in the table and close the excel sheet by clicking on **Exit** option from the **File** menu. The design table will be created and **Routing Library Manager** dialog box will be displayed again.

- Click on the **Next** button from the dialog box. The **Component Attributes** page of the dialog box will be displayed; refer to Figure-23.

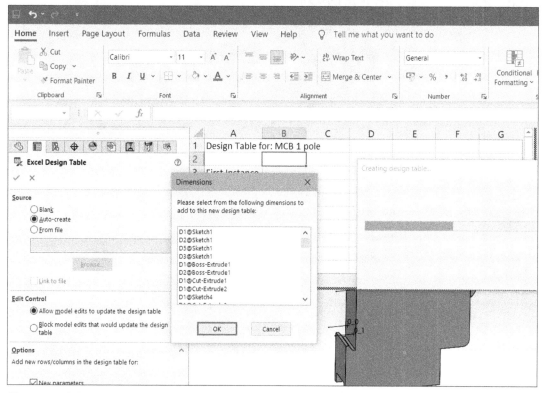

Figure-22. Selecting dimensions for design table

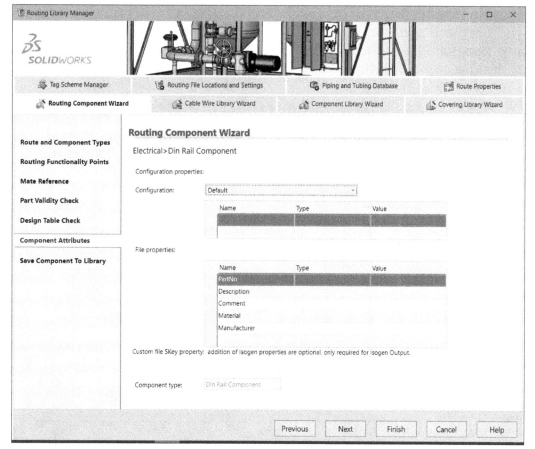

Figure-23. Component Attributes page

- Double-click in the fields and define desired parameters. After setting parameters, click on the **Next** button. The **Save Component To Library** page will be displayed in the dialog box; refer to Figure-24.

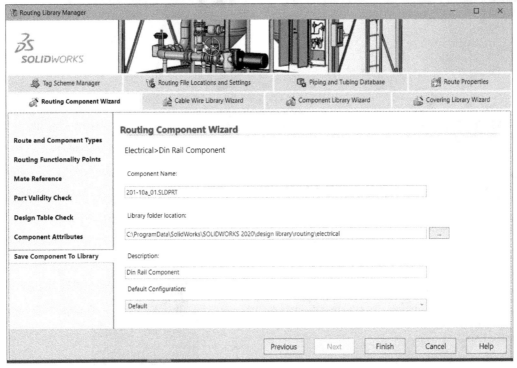

Figure-24. Save Component To Library page

- Specify the parameters as desired and click on the **Finish** button from the dialog box. An information box will be displayed. Click on the **Yes** button from the box to save the part and click on the **OK** button from next information box. The **Routing Library Manager** dialog box will be displayed again. Click on the **Cancel** button to exit the dialog box and close the dialog box.

DOWNLOAD SOLIDWORKS 3D PART

There is a big database of CAD parts from their manufacturers in electrical engineering available on internet. SolidWorks Electrical avails a tool named **Download SOLIDWORKS 3D Part** to use these manufacturer part files. The procedure to use this tool is given next.

- Click on the **Download SOLIDWORKS 3D Part** tool from the **Tools > SolidWorks Electrical** menu in the Menu bar; refer to Figure-25. You will be asked to specify login id and password. Register if not registered earlier and then login. The **CAD File Downloader** dialog box will be displayed; refer to Figure-26.

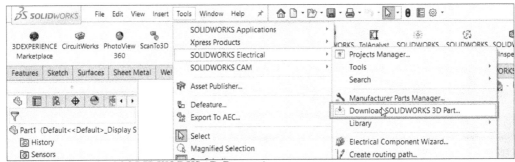

Figure-25. Download SOLIDWORKS 3D Part option

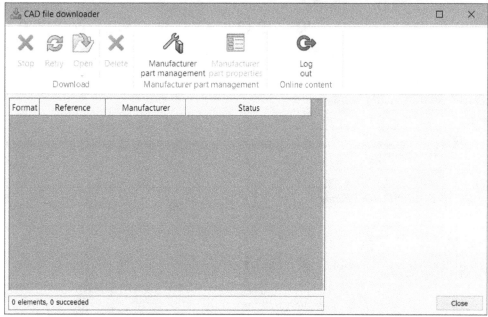

Figure-26. Cad File downloader dialog box

• Click on the **Manufacturer part management** button from the **Manufacturer parts management** panel in the dialog box. The **Manufacturer part management** dialog box will be displayed; refer to Figure-27.

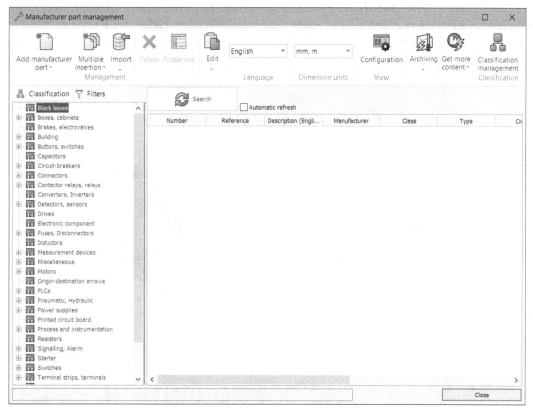

Figure-27. Manufacturer parts manager

• Select desired category from the left in the dialog box and search the component for which you want to download the part file.

• Select the component from the table in the right and click on the **Download SOLIDWORKS 3D part** button to download the part; refer to Figure-28. The **CAD file downloader** dialog box will be displayed and SolidWorks will start looking for the part over internet. Once the part is found, it will start downloading the part file; refer to Figure-29.

Figure-28. 3D SOLIDWORKS button

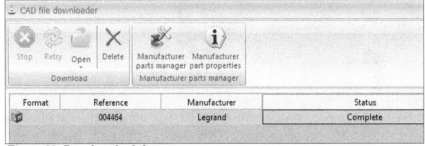

Figure-29. Part downloaded

- Select the part from the dialog box and click on the **Open** button from **Download** panel. Now, you can check the part in SolidWorks; refer to Figure-30. Note that you might be asked to proceed with feature recognition, choose **No** in such cases because FeatureWorks cannot identify mate references of SolidWorks Electrical.

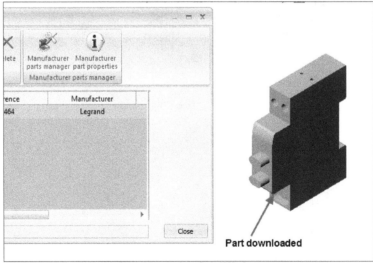

Figure-30. Downloaded part opened in SolidWorks

INSERTING COMPONENTS IN SOLIDWORKS ELECTRICAL 3D

In SolidWorks, you can insert the electrical components in the same way as you do in SolidWorks Assembly environment. But, there is also a special way designed to insert electrical components by using SolidWorks Electrical 2D projects. Before that we need to open the electrical project file in SolidWorks. The open project file is given next.

- In SolidWorks Electrical, open desired project file and click on the **SolidWorks assembly** tool from the **Processes** panel of the **Process** tab in the **Ribbon**. The **Creation of assembly files** dialog box will be displayed; refer to Figure-31.

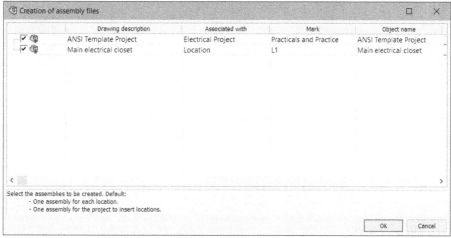

Figure-31. Creation of assembly files dialog box

- Select the check boxes for assembly files you want to add in project and click on the **OK** button. Blank assembly files will be added to the project.
- Start SolidWorks and add the **SolidWorks Electrical** add-in by using **Add-Ins** dialog box as discussed earlier.
- Click on **Electrical Project Management** tool from the **Tools > SolidWorks Electrical** menu; refer to Figure-32. The **Electrical Project Management** dialog box will be displayed; refer to Figure-33.

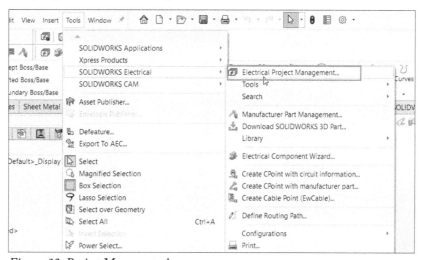

Figure-32. Project Manager tool

- Double-click on the project using which you want to create 3D panel drawing. The project file will open in **Electrical Project Pages** task pane at the right in SolidWorks; refer to Figure-34. Make sure that you have added SolidWorks cabinet layout file in the project by using SolidWorks Electrical. If you have not added 3D electrical assembly file then click on the **SOLIDWORKS Assembly** tool from the **Tools > SOLIDWORKS Electrical > Process** menu; refer to Figure-35.
- Expand the nodes of project and double-click on the cabinet layout part file; refer to Figure-36. An assembly file will open and components of the project will be displayed in **Component explorer** at the right of SolidWorks; refer to Figure-37.

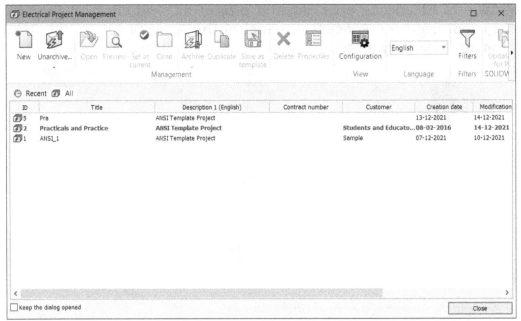

Figure-33. Project Manager dialog box

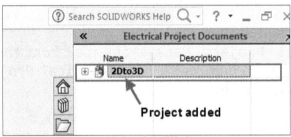

Figure-34. Project added in SolidWorks

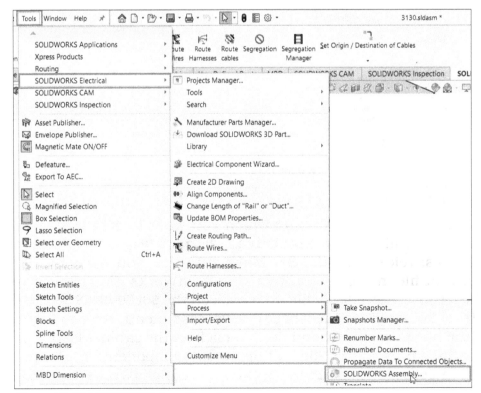

Figure-35. SOLIDWORKS Assembly tool

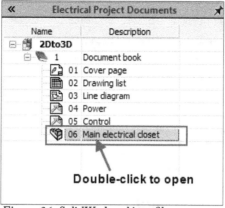

Figure-36. SolidWorks cabinet file

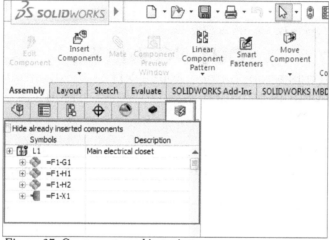

Figure-37. Components used in project

- Expand the node for a component in the **Component explorer** and double-click on the part file. The component will get attached to cursor; refer to Figure-38. In some cases, you may get a warning box like the one shown in Figure-39. In such cases, click on the **Insert the part** button from the box. If you do not want this warning box to be displayed then specify the size of component while defining its schematic/line diagram symbol.

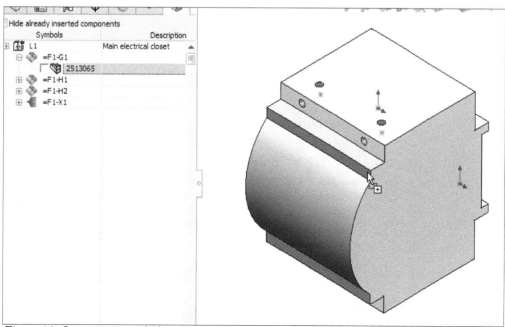

Figure-38. Component attached to cursor

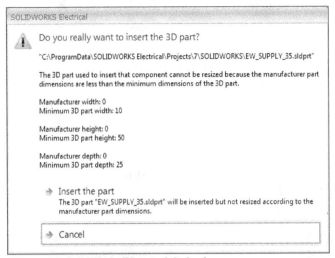

Figure-39. SolidWorks Electrical dialog box

ROUTING WIRES

SolidWorks Electrical has a tool named **Route Wires** to automate the process of wire routing based on the details provided in SolidWorks Electrical 2D. The procedure to use this tool is given next.

* Click on the **Route Wires** tool from the **SOLIDWORKS Electrical 3D** tab in the **Ribbon**; refer to Figure-40. The **Route wires PropertyManager** will be displayed; refer to Figure-41.

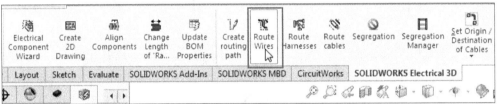

Figure-40. Route Wires

Figure-41. Route wires PropertyManager

- Select the **Show errors** check box from the **Routing Analysis** rollout in the **PropertyManager** if you want to check the errors generated while routing.
- Select desired option from the **Select route type** rollout. If you want to create wires that look alike real then select the **SOLIDWORKS Route** radio button. If you want to generate splines of different color representing wires then select the **3DSketch Route** radio button. Note that selecting the **3DSketch Route** radio button greatly reduces the time required by system for performing automatic routing.
- Select the **Cable core follow routing path** check box if you want to make the cores of cable follow the routing path as their wires do.
- There are two options to use as renderer for routing, **Use splines** and **Use lines**. Select desired option from the **Select renderer type** rollout. You can add tangency conditions at the links by selecting **Add tangency** check box from the rollout.
- Specify the routing parameters like deviation tolerance for distance between two routing paths or distance between connecting point and nearest routing path. These parameters are specified in the **Routing parameters** rollout.
- You can check the effect of specified parameters by selecting **Draw Graph** button; refer to Figure-42. Select the **Delete Graph** button to delete the graph.

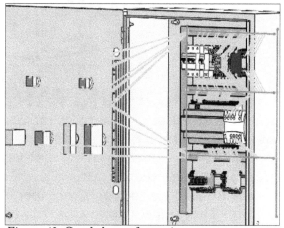

Figure-42. Graph drawn for routing

- From desired algorithm from the **Shortest Path Algorithm** rollout and selected desired engine from the drop-down displayed on selecting the algorithm.
- At last, click on the **OK** button from the **PropertyManager**. SolidWorks will start automatic routing and once the routing is complete, it will show errors/warnings of the routing via **Routing Analysis** window; refer to Figure-43.

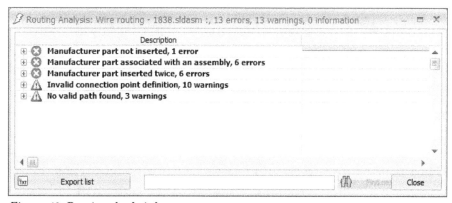

Figure-43. Routing Analysis box

- Expand each node and check the cause of error. Most of the time, these are simple errors like manufacturer part not inserted. In such cases, you need to insert the part. In some cases, you do not need a part to be inserted then you can ignore these errors. After performing modifications as per the errors, click again on the **Route wire** tool and perform routing.

Figure-44 shows an assembly after performing routing.

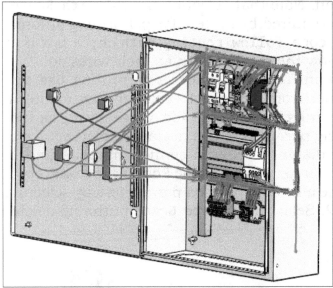

Figure-44. Assembly after routing

CREATE ROUTING PATH

The **Define Routing Path** tool is used to create path for routing wires. If we check the routing of assembly in Figure-44 then we can find that it is penetrating through the enclosure which is not generally considered in real-world; refer to Figure-45. To eradicate such situations, we use create routing path which tells system how to perform routing then wires should go through the specified path before making connection to the components. The procedure to use this tool is given next.

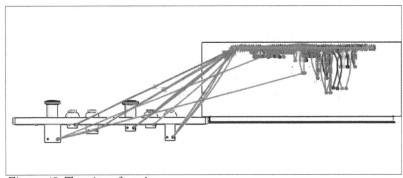

Figure-45. Top view of routing

- Click on the **Define Routing Path** tool from the **SOLIDWORKS Electrical 3D** tab in the **Ribbon**. The **Define Routing Path PropertyManager** will be displayed; refer to Figure-46.
- If you have already created 3D sketch for path in modeling area then select **Convert sketch** radio button from the **PropertyManager** and select the sketch. Click on the **OK** button from the **PropertyManager**. The routing path will be created.
- If you want to create a new routing path or you do not have an existing 3D sketch path then click on the **Create sketch** radio button and click on the **OK** button from the **PropertyManager**. A message box will be displayed telling you that only lines and sketch points can be used for creating routing path; refer to Figure-47.

Figure-46. Define Routing Path PropertyManager

Figure-47. SolidWorks Electrical message box

- Click on the **OK** button from the message box and draw the 3D sketch; refer to Figure-48.

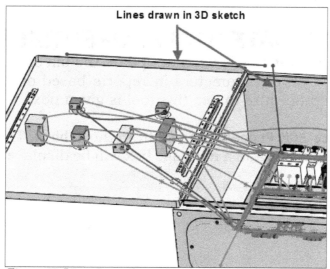

Figure-48. Lines drawn for routing path

- Exit the sketch environment. The routing path will be displayed in yellow color, by default.
- Now, click on the **Route Wires** tool from the **Ribbon** to check the difference caused in routing due to routing path; refer to Figure-49.

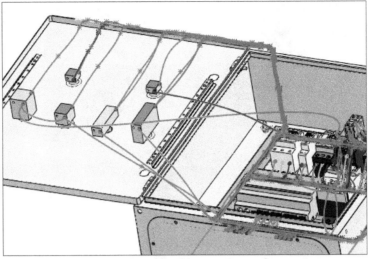

Figure-49. Modified routing after adding routing path

Note that you can add any component in the assembly as base for electrical routing. For example, you can add frame of a motorbike and place the electrical components on it before creating routing.

ROUTING CABLES AND HARNESSES

The **Route Cables** and **Route Harnesses** tools work in the same way as discussed for **Route wires** tool. Most of the options are same. There is a new option named **Update Origin/Destination** check box in case of routing cables which enables to update the origin and destination data of cable in the 2D schematics based on 3D routing and vice-versa.

UPDATE BOM PROPERTIES

The **Update BOM Properties** tool is used to update the bill of material by adding the cable/wire length and other parameters in reports based on the properties in 3D electrical model. The procedure to use this tool is given next.

*   Click on the **Update BOM Properties** tool from the **Ribbon**. The Bill of Material will be updated automatically and a message box will be displayed; refer to Figure-50.

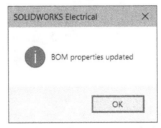

Figure-50. Message box stating updation of BOM

*   Click on the **OK** button to exit.

You can use the **Align Components** tool to align components and **Change Length of "Rail" or "Duct"** tool to change the length of rails and ducts in the assembly, as discussed earlier. The tools like Create 2D Drawing from assembly are native tools of SolidWorks. You can find the details of SolidWorks modeling and assembly tools in our other book, SolidWorks 2023 Black Book.

Annexure I

Basics of Electrical System

This chapter is an introduction to electrical system so we have given just brief introduction here. To learn more about electrical systems, you can check YouTube channel of **Mr. Jim Pytel** | Electro-Mechanical Technology Instructor | Columbia Gorge Community College at **https://www.youtube.com/user/bigbadtech** link.

Topics Covered

The major topics covered in this chapter are:

- *Basics of Electrical Circuits*
- *Types of Electricity*
- *Resistance Coding and Calculation*
- *Star & Delta Connections*
- *Voltage and Current Source*
- *Kirchhoff's Laws*
- *DC Circuits*
- *Electromagnetism*
- *Alternating Current*
- *Single AC Circuits*
- *Three Phase AC circuits*
- *Balanced and Unbalanced Circuits*

INTRODUCTION

In the previous part of this book, you have learned about AutoCAD Electrical commands and tools to create various schematics. This part of the book is about electrical engineering concepts that are important to understand before you start designing electrical panels. Depending on your need, you can start with this Part-II first and then you can work with Part-I of the book. We will start with basic terminology of electrical engineering.

WHAT IS ELECTRICITY?

Electrical engineering revolves around the fundamentals of electricity. It is concerned about how electricity is produced, transmitted, and manipulated to run various machines. Here, the basic question is what is electricity? Electricity is a form of energy caused by the presence or flow of electrons. If free electrons are present in a metal in charges state then they produce static current. If the electrons are flowing from one point to another point in a metal conductor then it is called live current. So, current is one of the parameter used to measure electricity. There are few more parameters used to measure electricity which are discussed next.

How Electricity is Measured?

Every material around us is made up of elements. Every element is made up of atoms and every atom is composed of electrons, protons and neutrons (At least till writing this book!!). The electrons have negative charge, the protons have equal positive charge, and neutrons as name suggests are neutral. Due to external influence, an electron can be freed from its atom and forced to move in desired direction. The charge induced by one such electron is measures as 1.602×10^{-19} C (coulomb). So, we can say that the quantity by which we can measure electricity is **coulomb**.

Current

The coulomb value of a metal is not feasible to measure as there can be millions of electrons in small section of metal and we might need supercomputers to measure the charge in this small section. So, we derived a smart parameter to measure charge called **current**. Current can be defined as rate of flow of charge particle through a point on the conductor (A conductor is a material which allow the current to pass through due to availability of free electrons). In mathematical terms, it can be defined as :

$$I = \frac{dq}{dt} = -nev_d$$

Here, I is current, dq/dt is the rate of charge per second
n is the number of electrons, -e is the charge on electrons, and vd is drift velocity of electron.

The unit of current is ampere denoted by A. One ampere current is the flow of 1 C (coulomb) charge from a conductor at a point per second.

There are two types of current; direct current (DC) and alternating current (AC). Graphically, we can define these currents as shown in Figure-1.

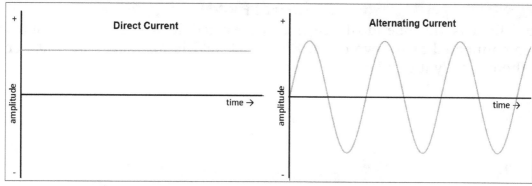

Figure-1. Types of currents

As you can see from the figure, in direct current, the amplitude of current remains constant with respect to time and flow is unidirectional. In case of alternating current, the amplitude of current changes both value as well as direction. Generally, AC current follows two wave forms, sinusoidal wave and square wave. There can be other forms of AC currents but we will not delve deeper here.

Current Density

The current density if the amount of current carried by conductor per unit area. The area should be perpendicular to the direction of current flow. Mathematically, it can be given by formula:

$$J = \frac{I}{A}$$

Since, current density is dependent of current flow direction so it is a vector quantity and its unit is A/m².

Voltage or Potential Difference

The **Voltage** or potential difference is the amount of work required to move per coulomb charge from one point to another point in conductor. Unit of voltage is Volt or simply V. Mathematically, it can be given as:

$$V = \frac{W}{Q}$$

So, 1 volt = 1 Joule/1 Coulomb.

When representing voltage in electrical drawing, it should always be defined with + and - signs as shown Figure-2.

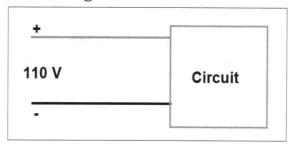

Figure-2. Voltage representation

Electrical Field

Electrical field is the pseudo force field generated in a conductor when voltage source is connected at its two ends. Electrical field is measures by volt per meter and mathematically it can be defined as:

$$E = \frac{V}{L}$$

Electrical Power

The electrical power is defined as product of Voltage and current, P = V x I. The unit of power is watt (W). You can also define power in horsepower (hp). One hp is equal to 746 W.

Efficiency

The electrical efficiency of a component, circuit, or system can be defined as useful power output divided by total power consumed. The efficiency is generally denoted in percentage (%). Mathematically, it can be given as:

$$Efficiency\ (\%) = \frac{Useful\ Power\ Output}{Total\ Power\ Consumed}$$

OHM'S LAW

Ohm's law is the most basic rule in electrical engineering for circuits. As per the Ohm's law, the Voltage across two points of a conductor is directly proportional to the current flowing through it.

$$V \propto I$$

Or, say V = R x I where R is constant of proportionality (Also called Resistance)

Resistance

The resistance to flow of current in a material is called Resistance. The unit of resistance is ohm (Ω). Mathematically, R is calculated by:

$$R = \rho \frac{L}{A}$$

Here, ρ is resistivity defined by unit ohm meter.
       L is the length of conductor
       A is the cross-section area of conductor.

There are mainly two types of resistors you will come across during practical use; wire wound resistors and carbon molded resistors. There is also another category of resistors called metal film resistors. The metal film resistors have very low power rating and can be easily damaged due to power surge.

Wire Wound Resistors

The wire wound resistors are those in which insulated resistance wire is wound around heat-resistant ceramic tubes; refer to Figure-3. These resistors can be of fixed resistance or variable resistance. The resistance value is based on the length of wire and its resistivity. The value of resistance goes from fraction of ohm to kilo ohms. You can find the resistance value and rated power value printed on the resistors. The power rating of these resistors can be from few watts to 1000 watts.

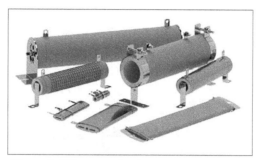

Figure-3. Wire wound resistors

Carbon Molded Resistors

In electronic circuits and other precise electrical circuits where high resistance is required, the carbon molded resistors are the best solution. The cost of creating wire wound resistor for mega ohm resistance will be very high whereas comparatively, you can get a carbon molded resistor of same rating at very low cost. Note that although resistance value of carbon molded resistors is high but their power rating is generally low (in the range of 1/4 W to 5 W). The carbon molded resistors follow a color coding to denote their resistance value; refer to Figure-4. (Want to learn more follow the link https://www.wikihow.com/Identify-Resistors). Note that metal film resistor might look similar to carbon molded resistors but you can identify the two by their background color. The metal film resistors generally have blue background color whereas carbon molded resistors have light brown color background.

Color	Signficant figures			Multiply	Tolerance (%)	Temp. Coeff. (ppm/K)	Fail Rate (%)
black	0	0	0	x 1		250 (U)	
brown	1	1	1	x 10	1 (F)	100 (S)	1
red	2	2	2	x 100	2 (G)	50 (R)	0.1
orange	3	3	3	x 1K		15 (P)	0.01
yellow	4	4	4	x 10K		25 (Q)	0.001
green	5	5	5	x 100K	0.5 (D)	20 (Z)	
blue	6	6	6	x 1M	0.25 (C)	10 (Z)	
violet	7	7	7	x 10M	0.1 (B)	5 (M)	
grey	8	8	8	x 100M	0.05 (A)	1(K)	
white	9	9	9	x 1G			
gold			3th digit only for 5 and 6 bands	x 0.1	5 (J)		
silver				x 0.01	10 (K)		
none					20 (M)		

6 band — 3.21kΩ 1% 50ppm/K

5 band — 521Ω 1%

4 band — 82kΩ 5%

3 band — 330Ω 20%

gap between band 3 and 4 indicates reading direction

Figure-4. Resistor color codes chart

Most of the time, resistance is intentional in circuit to achieve desired objectives like reducing current, generating heat energy, and so on.

Series and Parallel Connections

Based on the application, you can connect multiple resistors in a circuit in series, parallel or combination of both; refer to Figure-5.

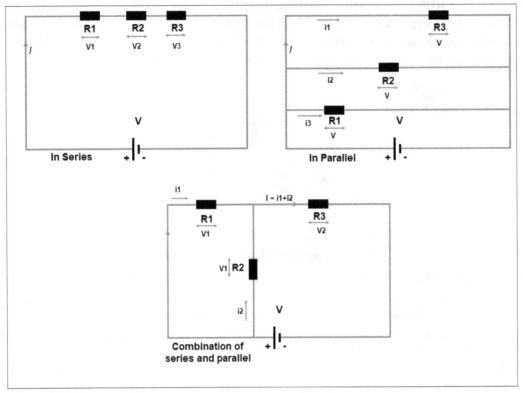

Figure-5. Resistor connections

Note that when we connect resistors in series in a circuit then the total voltage is divided among the components but current flowing through them is same.

Using this information in equation, we get

$$V = V1+V2+V3$$

$$IR = IR1+IR2+IR3 = I(R1+R2+R3)$$

Total Resistance $R = (R1+R2+R3);$

So, we can say that when resistors are connected in series then we can add all the values of resistors to get equivalent resistance value.

When we connect resistors in parallel in a circuit then voltage across each resistor is same but current gets divided among the components so we can say that,

Total current $I = I1+I2+I3$
$V/R = V/R1+V/R2+V/R3 = V(1/R1+1/R2+1/R3)$
$1/R = 1/R1+1/R2+1/R3$

Star and Delta Connections

While working with circuits, you will not find all the resistors in series or parallel, there can be some devil configurations as shown in Figure-6.

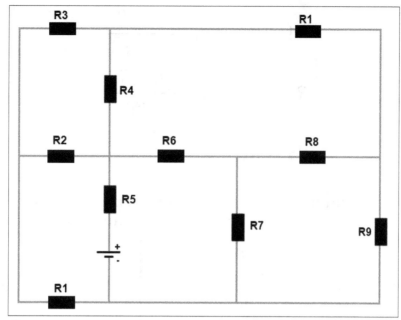

Figure-6. Complex connections of resistors

In such cases, we are looking at delta connections like R7, R8, and R9 form one delta connection and R2, R3, and R4 form another delta connection. There resistors are neither in parallel and not in series. The solutions to such problems are star-delta and delta-star transformation equations; refer to Figure-7.

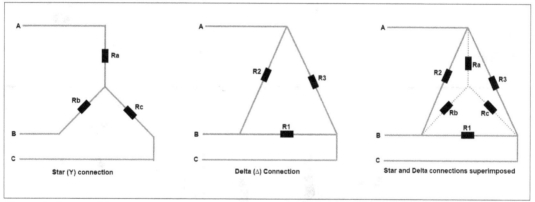

Figure-7. Star and Delta connections

For Delta to Star Transformation, the values are :

$$Ra = \frac{R2 \cdot R3}{R1 + R2 + R3}$$

$$Rb = \frac{R3 \cdot R1}{R1 + R2 + R3}$$

$$Rc = \frac{R1 \cdot R2}{R1 + R2 + R3}$$

For Star to Delta transformation, the values are:

$$R1 = Rb + Rc + \frac{Rb \cdot Rc}{Ra}$$

$$R2 = Rc + Ra + \frac{Rc \cdot Ra}{Rb}$$

$$R3 = Ra + Rb + \frac{Ra \cdot Rb}{Rc}$$

Now, if we apply the delta to star transformation then problem shown in Figure-6 can be reduced to solution show in Figure-8 which can be solved easily.

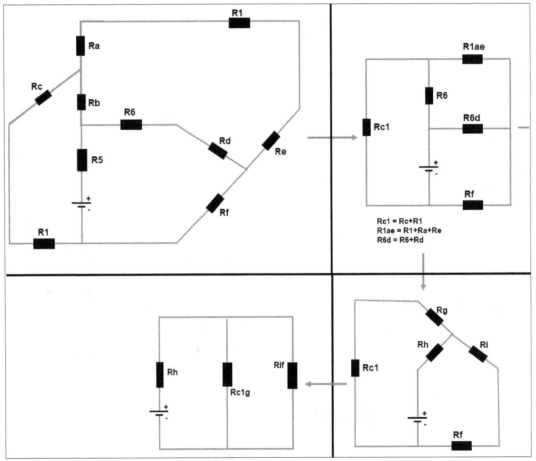

Figure-8. Solution approach for complex connection

CAPACITORS AND INDUCTORS

Capacitors and Inductors, both draw power from the source, store it and then discharge it when there is a sudden change in current/voltage in the circuit. Both the components are widely used in filtering signals and AC circuits. The main difference between the two is how they store the energy. A capacitor stores the energy in the form of electrical field in a dielectric material trapped between two conductor plates. However, an inductor stores the energy in the form of magnetic field in coils and loops. Also, a capacitor activates on change in potential difference whereas inductor acts when there is a change in current. The unit of capacitance is farad (F) which describes the amount of current (in ampere seconds) a capacitor can store/discharge per volt change. The symbol for capacitor is C in a schematic. The unit of inductance is henry (H) which describes the amount of voltage (in volt seconds) it can provide per ampere in case of change in current of circuit. The symbol for inductor is L in a schematic. You can easily identify capacitors and inductors by looking at them; refer to Figure-9.

Figure-9. Capacitors and Inductors

Inductors in Series and Parallel

The inductors combine in the same way as resistors do in series and parallel.

In series:

$$L = L_1 + L_2 + L_3 + \ldots\ldots\ldots\ldots$$

In parallel:

$$1/L = 1/L_1 + 1/L_2 + 1/L_3 + \ldots\ldots\ldots\ldots$$

Capacitors in Series and Parallel

The capacitors combine in the way opposite to resistors do.

In series:

$$1/C = 1/C_1 + 1/C_2 + 1/C_3 + \ldots\ldots\ldots\ldots$$

In Parallel:

$$L = C_1 + C_2 + C_3 + \ldots\ldots\ldots\ldots$$

COMBINATION OF ENERGY SOURCES

The are mainly two energy sources in electrical circuits; voltage source and current source. For simplification of circuits when there are two voltage sources in series then we can sum their values to express total voltage; refer to Figure-10. Note that you can combine two voltage sources (mainly batteries) in parallel if both the sources have same voltage values to increase total current storage capacity but you cannot combine sources with different voltage values in parallel as they will cause short circuit in battery of lower voltage. Similarly, when there are two current sources in parallel then we can sum their values to express total current; refer to Figure-11. Note that current sources should not be combined in series because no one can predict what will be the output current, it can be combination of both current value, it can be value of one current source. Keep in mind that practically batteries are voltage sources not the current sources.

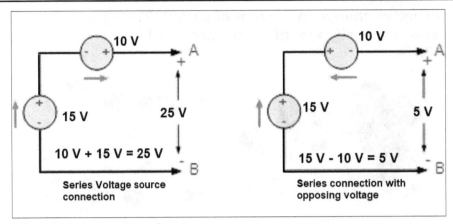

Figure-10. Voltage source combinations

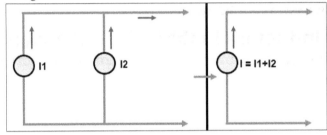

Figure-11. Current source combination

KIRCHHOFF'S LAWS

Gustav Robert Kirchhoff stated two laws for circuit equations which can be used to determine current and voltage values at any section of circuit. These two laws are called Kirchhoff's Current Law (KCL) and Kirchhoff's Voltage Law (KVL).

The Kirchhoff's Current Law (KCL) states that the algebraic sum of current incoming to a point in circuit and the current outgoing from that point in circuit is zero; refer to Figure-12.

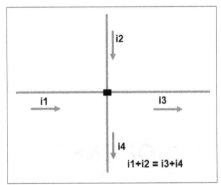

Figure-12. Kirchhoff's Current Law

The Kirchhoff's Voltage Law (KVL) states that the algebraic sum of voltages around a closed loop in circuit is zero; refer to Figure-13. Note that when you encounter + sign before a voltage source then you need to add it in total voltage and when you encounter - sign before a voltage source then you need to subtract it from total voltage in the loop.

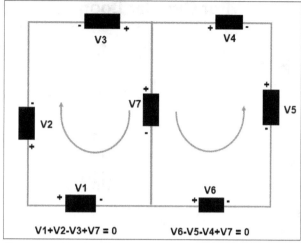

$$V1+V2-V3+V7 = 0 \qquad V6-V5-V4+V7 = 0$$

Figure-13. Kirchhoff's Voltage Law

ELECTROMAGNETISM

When direct current passes through a wire then a magnetic field is setup in its vicinity in the form of concentric circles. The direction of magnetic field can expressed by right-hand thumb rule which goes like stretch the thumb of your right hand along the current. The curl of fingers will give the direction of magnetic field; refer to Figure-14.

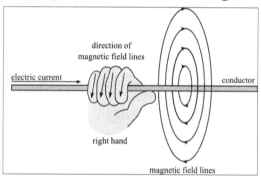

Figure-14. Thumb rule for electromagnetism

Magnetic Field Strength

The strength of magnetic field at a point on wire is directly proportional to current and inversely proportional to its distance from start point. The magnetic field strength parameter is denoted by B and its unit is tesla (T). The formulas for magnetic field strength of different wiring arrangements are given next.

For straight wire

$$B = \frac{\mu_0 I}{2\pi x}$$

Here, I is current, x is distance from start point and μ_0 is permeability of air which has a value of $4\pi \times 10^{-7}$ Tm/A.

For Circular Loop

$$B = \frac{\mu_0 I}{2r}$$

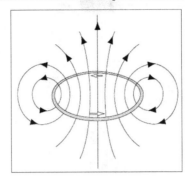

Here, r is the radius of circular loop.

For Solenoid

At center of solenoid, the magnetic field strength is given by,

$$B = \mu_0 nI$$

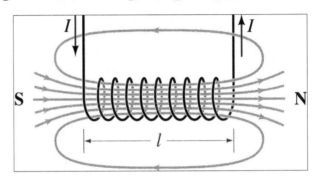

Here, n is the number of turns per unit length

At the either end of solenoid, the magnetic field strength gets halved.

For Toroid

$$B = \frac{\mu_0 nI}{2\pi r}$$

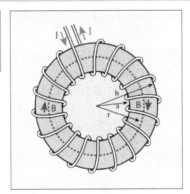

ALTERNATING CURRENT

Alternating means changing direction so the current and voltage that changes direction is called alternating current and alternating voltage, respectively. In alternating current supply, the current varies sinusoidally and can be mathematically represented by;

$$i = I_m \sin\omega t$$

which can be graphically represented as:

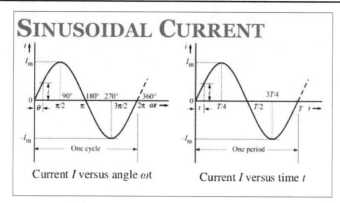

Common Terms for ac supply:

Frequency : Frequency of alternating current is the number of cycles that occur per second. In simple words, it is the number of times current value becomes positive in a second. The unit for frequency is Hz (hertz) which is cycles per second. Frequency is denoted by *f*. In USA, the frequency of alternating current supply is 60 Hz.

Cycle : One complete set of positive and negative values of waveform is called cycle.

Peak or Maximum Value : It is the maximum value of current generated by waveform. It is also called **amplitude** and denoted by I_m.

Time Period : It is the time in second required to complete one cycle. It is denoted by T and is reciprocal of Frequency (F).

Angular Frequency : It is the number of radians covered per second. Angular frequency is denoted by ω which equal to 2π*f*.

Phase : Phase is used to identify if there a mismatch in two waveforms of alternating current and if there is a mismatch then what is its value. Figure-15 shows two alternating current that are in same phase as the two currents attain their maximum, minimum, and zero values at the same time. If there is a difference between two waveforms then the difference is called phase difference or phase shift.

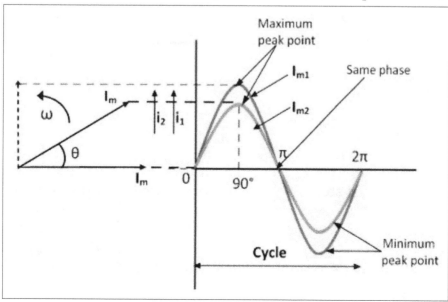

Figure-15. Phase of alternating current

Three Phase Voltage

In a three phase voltage system, three emf of same magnitude and frequency are applied which are 120 degree phase displaced to each other; refer to Figure-16. The line voltage for three phase supply is 440 V (i.e. when you use two live lines for connection) and phase voltage for three phase supply is 220 V (i.e. voltage between neutral and any one live line).

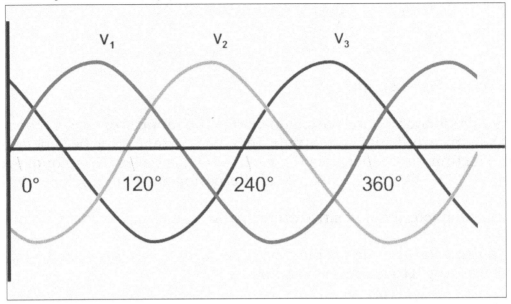

Figure-16. Three phase voltage waveform

Three Phase Loads

The three phase loads are connected to the three phase supply in either Star form or Delta form; refer to Figure-17.

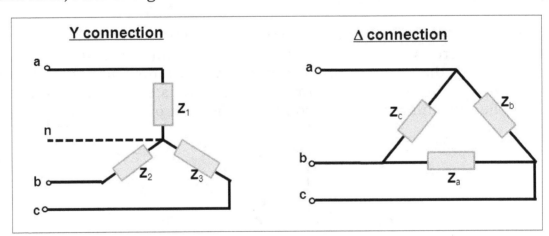

Figure-17. Three phase load system

In a three phase load system if all the three impedances (loads) equal then it is called balanced load system. If the three loads are of different values then it is called unbalanced load system. Generally, three phase supply is provided for both industrial and domestic use. In industries, most of the machines run on balanced load but in domestic application, the loads are not always balanced on supply lines. If an unbalanced load is connected in star configuration to the supply then voltage drops for different branches will be different causing situation called neutral shift.

In such cases, voltage in one line can be significantly high whereas in another line it might be very low. So, it is suggested to avoid unbalanced load and if it is not possible then add pseudo loads so that circuit gets balanced. If possible use 3 line + 1 neutral configuration for three phase system as extra current will pass through neutral line and will keep the system balanced at cost of some extra energy.

FOR STUDENT NOTES

Annexure II

Basics Components of Electrical Control System

Topics Covered

The major topics covered in this chapter are:

- *Switches*
- *Relays and Timer Relays*
- *Contactors*
- *Fuses, Circuit Breakers and Disconnectors*
- *Terminals and Connectors*
- *Solenoids*
- *Motors and Transformers*
- *Sensors and Actuators*

INTRODUCTION

There are various components involved in designing an electrical control circuit like push buttons, relays, contactors, disconnectors and so on. When designing an electrical control system, it is important to understand the parameters of these components and how they fit in your circuit. In this chapter, we will learn about some common electrical components used in these circuits.

PUSH BUTTONS

The push buttons are used to start and stop motors and other electrical machines. There are mainly two type of push buttons based on operation; Normally Open (NO) and Normally Closed (NO); refer to Figure-1. These push buttons are generally coupled with contacts to run motor or other electrical machines. So, once you have pressed NO push button to start a circuit, the contact will keep on supplying current until NC push button is pressed; refer to Figure-2. The contact points in good quality push buttons are made of silver to provide long life and better conductivity. Gold contacts are typically necessary when switching at logic level, generally defined as covering 1 to 100 mA. The momentary voltage on push buttons is very high so you should make sure that proper insulation is provided in the push button depending on your application. The factors involved in selection of push button are load, voltage, contact materials, circuit type, terminal type, and mounting.

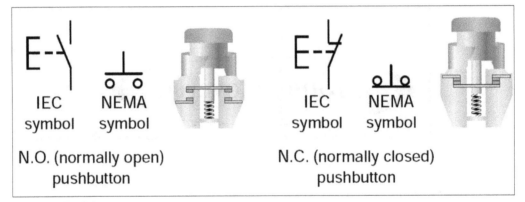

Figure-1. Push buttons

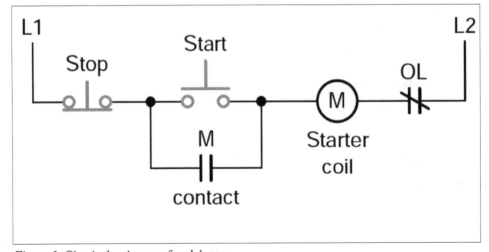

Figure-2. Circuit showing use of push button

PILOT LIGHTS

The pilot lights provide visual indication of circuit status. If a pilot light is glowing then it means the circuit is connected and power is running through it. If you have controls install at one place and connected motors are half a mile away then there is no way to instantaneously confirm whether motor is running or not after you have pressed the push button for it. In such case, pilot lights are connected in motor circuit. The pilot lights are available in different colors which are mainly Red, Yellow, Green, Blue, White, Grey, and Black. As per IEC standards, the use of these lights is described in Figure-3.

Pilot Light Color and its meaning		
Color	Meaning	Use
RED	Emergency	It is used to show emergency stop of a machine
YELLOW	Abnormal condition	It is used to show there is some problem in circuit
GREEN	Normal	It is used to show the circuit is working properly
BLUE	Action required	It is used to show that manual interventaion is required for machine
WHITE, GREY, BLACK, CLEAR	Neutral	These lights are used for custom indications for example if a specific part of machine is to be monitored then assign a pilot light to it.

Figure-3. Pilot color coding

Legend Plate

Legend plate is installed around push buttons and pilot lights to identify their purpose. The legend plate text includes text like START, STOP, RESET, RUN, UP, DOWN, and so on.

SWITCHES

Various types of switches are used in electrical systems like selector switch, toggle switch, drum switch, limit switch, temperature switch, pressure switch, float & flow switches, and so on. These switches are discussed next.

Selector Switch

The selector switches are used where you need to select from a broad range of contacts; refer to Figure-4. The selector switch use rotating cam for their operation. You can easily find the use of selector switch in washing machines, dishwashers and other home appliances where you need to select a mode.

Figure-4. Selector switch

The selector switches can be illuminated, non-illuminated, and non-illuminated with key. Selector switch with key are used when machine being operated by that switch can dangerous for people around it if not handled properly. So, the person authorized with key can only operate the machine. The specifications for selection of selector switch are lighted/non-lighted/key-type, physical dimensions, color coding, and current rating.

Toggle Switch

The toggle switch is used to manually toggle between On and Off position of the circuit. The switches on switchboards found in our home to on/off fans, lights, and other appliances are a type of toggle switches. There are mainly four types of toggle switches; SPST (Single Pole Single Throw), SPDT (Single Pole Double Throw), DPST (Double Pole Single Throw), and DPDT (Double Pole Double Throw); refer to Figure-5.

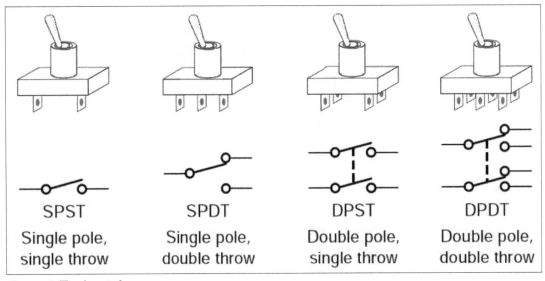

Figure-5. Toggle switches

Single pole single throw (SPST) toggle switches make or break the connection of a single conductor in a single branch circuit. This switch type typically has two terminals and is referred to as a single-pole switch.

Single pole double throw (SPDT) toggle switches make or break the connection of a single conductor with either of two other single conductors. These switches usually have three terminals and are commonly used in pairs. SPDT switches are sometimes called three-way switches.

Double pole single throw (DPST) toggle switches make or break the connection of two circuit conductors in a single branch circuit. They usually have four terminals.

Double pole double throw (DPDT) toggle switches make or break the connection of two conductors to two separate circuits. They usually have six terminals are available in both momentary and maintained contact versions.

Specifications for selection of toggle switch are dimensions, electrical ratings, terminal types, materials, and features.

Drum Switch

In drum switches, you can open or close contacts by moving center shaft; refer to Figure-6. The drum switches can be used to forward, reverse and stop the motor. The parameters for selection drum switch are number of poles, current rating, Switch type (reversing/non-reversing), voltage rating, AC/DC, and so on. You will find the connection settings for reversing motor in the manual supplied with the drum switch. The general schematic for forward and reverse operation by drum switch can be given as Figure-7.

Figure-6. Drum switch

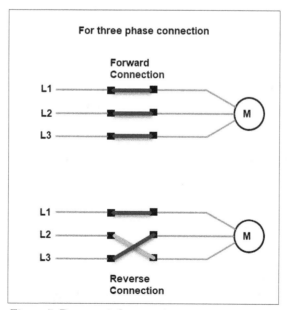

Figure-7. Drum switch connections

Limit Switch

The limit switches work automatically based on various factors like pressure, temperature, position and so on; refer to Figure-8. When the governing factor for limit switch reaches a certain threshold then the switch starts functioning. For example, there is a normally closed limit switch which will keep closed until the pressure reaches 30 bars inside a container. When pressure inside the container reaches 30 bar then the switch will automatically open. There are mainly two parts of a limit switch; body and actuator. The body part contains contacts that open or close based on movement of actuator. The symbols for limit switches are shown in Figure-9.

Figure-8. Limit switch

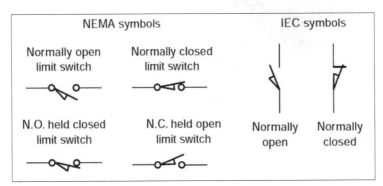

Figure-9. Limit switch symbols

Micro-limit switch is a variation of limit switches in which snap action switch is used to quickly transfer contact from one point to another point. Micro limit switches are available in single pole double throw configuration and one terminal for both NC and NO contacts is common.

Rotating cam limit switch is used where you want to control to rotation of a shaft in machine. They are useful to cutoff power at certain position or rpm threshold.

Selection of limit switch depends on type of control (rotary or linear), type of switch (electromechanical or solid state), contact types, poles and throw configuration, and so on.

Temperature Switches

Temperature switches are actuated by temperature or change in temperature. The symbol for temperature switch is given in Figure-10. The most common mechanism used in temperature switch is bimetallic strip which has two different metals having different thermal expansion properties joined together. Heating or cooling this bimetallic strip causes it to bend causing switch to close or open.

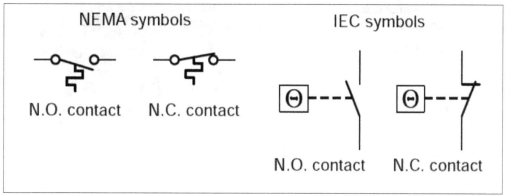

Figure-10. Temperature switch symbols

There is another type of thermal switch which works on expansion of fluid due to temperature called liquid filled temperature switch. It comprises a brass bulb filled with a chemical fluid (sometimes gas). It includes a small tube which hooks up the bulb to a pressure sensing mechanism consisting of bellows, bourdon tube or diaphragm. The functioning of both switches is shown in Figure-11.

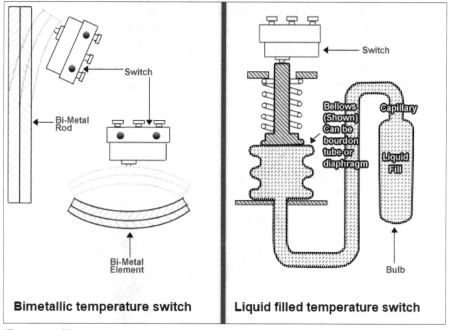

Figure-11. Temperature switches

The selection criteria for temperature switch includes temperature range, current and voltage rating, type of temperature switch required, and so on.

Pressure Switches

The pressure switch is used to control and monitor pressure. The most common application of pressure switch is found in air compressor system where the pressure switch causes to stop motor after pressure in the tank has reached a certain threshold. The symbols for pressure switches are given in Figure-12. The stiffness of spring defines the limit at which pressure switch will be activated.

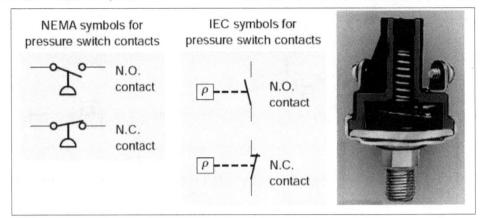

Figure-12. Pressure switch symbols

The selection criteria of pressure switch includes fixed or adjustable type pressure limit, type of fluid, fatigue, electrical output, pressure and electrical connections, response time, and ratings.

Liquid Level and Flow Switches

A liquid level switch is also called float switch as the bulb of switch floats on liquid. This switch is used when you want to cut off power once liquid in container has reached certain level. There is another variation of this switch in which two wires of low current are suspended in the liquid tank and when a conductive liquid submerges both the wire ends, the circuit gets closed causing the switch to activate. Both the types are shown in Figure-13. The symbols for float switch are shown in Figure-14.

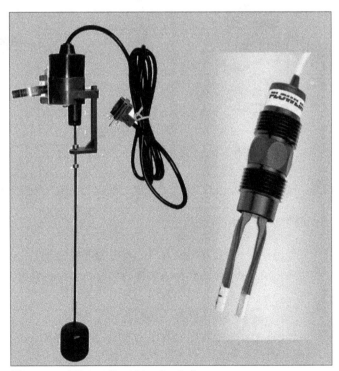

Figure-13. Float or liquid level switches

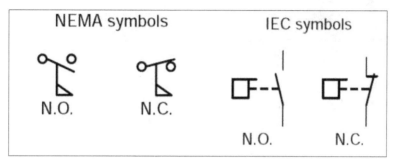

Figure-14. Symbols for float switch

The flow switches are used to monitor flow of fluid and control opening/closing of contacts accordingly. The symbols for flow switches are shown in Figure-15. Figure-16 shows different types of flow switches.

Figure-15. Symbols for flow switches

Figure-16. Flow switches

Joystick Switch

A joystick switch is used to control movement of a machine part in more than one direction. They are manually operated control switches and you can easily find their applications in backhoe loaders, cranes, remote control devices and so on. The symbol of joystick can be given as :

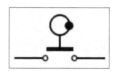

There can be upto 8 actuation directions with different switch combinations on a joystick; refer to Figure-17.

Figure-17. Joystick switches

The selection criteria for joystick includes number of contacts, current & voltage ratings, material of controller, contact types, and so on.

Proximity Switches

The proximity switches detect the presence of different shape, size, and type of materials based on sensors. Based on the data provided by sensors, the proximity switches activate the desired function. The sensors used in proximity switches can be different types like light sensor, pressure sensor, emf sensors, and so on. The symbols for proximity switches are shown in Figure-18.

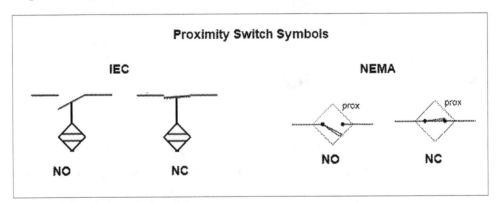

Figure-18. Proximity switch symbols

The point at which the proximity sensor confirms an incoming target is called the operating point. The point at which an outgoing target makes the switch to return to normal position is called the release point. Most proximity sensors also have LED indicator the status. The area between operating and release points is known as the hysteresis zone. The sensors in proximity switches generally work on either 24 V DC or 120 V AC power supply. Selection criteria for proximity switches include range of sensor, type of material to be sensed, space for sensor, environment, output type, contact type, and so on.

Sensors

Sensors are use to transmit digit or analog signals based on detected parameter. There are various types of sensors available for control designers like proximity sensor, temperature sensor, pressure sensor, humidity sensor, radiation sensor, and so on; refer to Figure-19. Since there is a long list of sensors now available in market, it is not feasible to discussed every sensor here. As an electrical control designer, your task will be to select appropriate sensor for your job. For example, you need to run a boom barrier motor at toll plaza based on information received through an id scanner. Now, you need to check the list of sensors and find out which one suits you the best. Like in this case, an RFID sensor will be best choice when each car has an RFID installed in it. The next parameters for electrical engineer will be voltage and current ratings of sensor, range of sensor, environmental conditions for sensor, contact types for sensor, and so on.

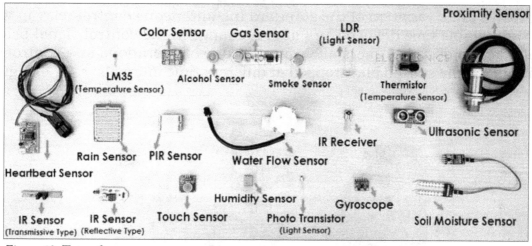

Figure-19. Types of sensors

ACTUATORS

Actuators are the electrical components that convert electrical energy to mechanical movement. There various types of actuators in electrical control design like relay, solenoids, and motors. The brief description of these actuators is given next.

Relays

A relay consists of two parts; relay coil and contact which are packed as a unit with separate contact points. When current is passed through relay coil then due to electromagnetism, it attracts iron arm connected to contact. The symbol for relay is shown in Figure-20. The working principle of relay is shown in Figure-21. So, whenever you need high power contacts controlled by low power switches then you can use relays.

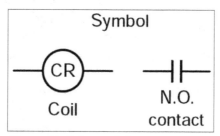

Figure-20. Relay symbol

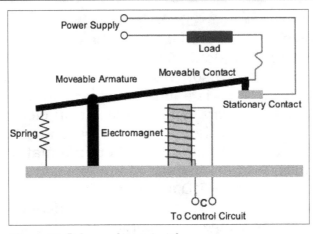

Figure-21. Relay working principle

Time Relays

Timing relays are a variation of the standard instantaneous control relay in which a fixed or adjustable time delay occurs after a change in the control signal before the switching action occurs. Timers allow a multitude of operations in a control circuit to be automatically started and stopped at different time intervals; refer to Figure-22.

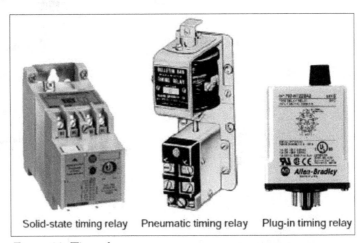

Solid-state timing relay   Pneumatic timing relay   Plug-in timing relay

Figure-22. Time relays

Latching Relays

Latching relays typically use a mechanical latch or permanent magnet to hold the contacts in their last energized position without the need for continued application of coil power. They are especially useful in applications where power must be conserved, such as a battery-operated device, or where it is desirable to have a relay stay in one position if power is interrupted.

Solid State Relays

A solid-state relay (SSR) is an electronic switch that, unlike an electromechanical relay, contains no moving parts. Although EMRs and solid-state relays are designed to perform similar functions, each accomplishes the final results in different ways. Unlike electromechanical relays, SSRs do not have actual coils and contacts. Instead, they use semiconductor switching devices such as bipolar transistors, MOSFETs, silicon-controlled rectifiers (SCRs), or triacs mounted on a printed circuit board. All SSRs are constructed to operate as two separate sections: input and output. The input side receives a voltage signal from the control circuit and the output side switches the load.

Solenoid

Solenoid is an electrical component in which uses electrical energy to move an armature electromagnetically. There are mainly two types of solenoids, linear and rotary with linear and rotary working directions respectively for mechanical output. The symbol for solenoid coil is shown in Figure-23.

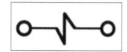

Figure-23. Symbol for solenoid coil

There are two power supply options for solenoid coils, AC and DC. The choice mainly depends on the type of supply available. If both the supplies are available then AC solenoids are more powerful in fully open position as compared to DC solenoids. AC operated solenoids are faster in response as compared to DC solenoids. The working principle of solenoid coil is shown in Figure-24.

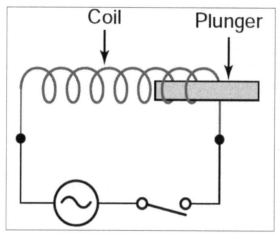

Figure-24. Solenoid coil working principle

Solenoid Valve

In solenoid valves, the opening and closing of valve is controlled by solenoid coil. These valves are mainly used to control opening and closing for hazardous fluids like nitrogen, gas, acids, and so on. The working principle of solenoid valve is shown in Figure-25. The symbols for solenoid valve are shown in Figure-26.

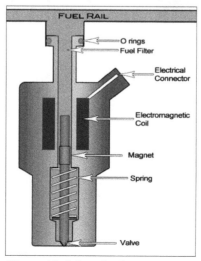

Figure-25. Solenoid valve

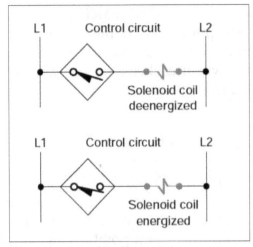

Figure-26. Solenoid valve symbols

MOTORS

Motors are also a type of actuators with rotary motion but since they are widely used in industries, it is better to discuss their types also. All motors work on the same electromagnetism principle: you pass electricity through a coil, it gets magnetized and starts to pull/push a mechanical plunger. In case of motors, there are two magnets instead of one. The same poles (N-N or S-S) of two magnets repel each-other while different poles (N-S) of two magnets attract each other. We will first discuss DC motors and then we will discuss with AC motors.

DC Motors

In DC motors, the current does not change direction and remains constant. The working principle of DC motor is shown in Figure-27.

Brushed Electrical DC Motor

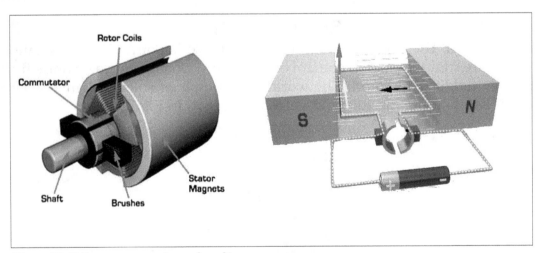

Figure-27. DC motor construction and working

There are mainly four types of brushed DC motors: permanent magnet dc motor, series dc motor, shunt dc motor, and compound dc motor. The applications of these motors are discussed next.

1. **Permanent Magnet DC Motor** : In this type of DC motor, permanent magnets of fixed capacity are placed for field magnets while rotor has coils for electromagnetism; refer to Figure-28. This type of DC motor provides great starting torque and has good speed regulation, but torque is limited so they are typically found on low horsepower applications.

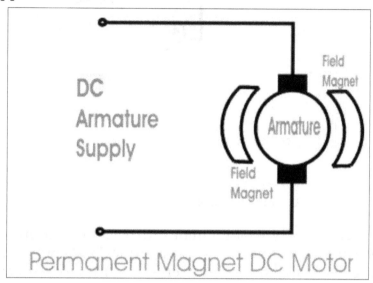

Figure-28. Permanent magnet dc motor

2. **Series DC Motor** : In this type of DC motor, the same current flows through rotor winding with is flowing through field winding; refer to Figure-29. The series DC motors create a large amount of starting torque, but cannot regulate speed and can even be damaged by running with no load. These limitations mean that they are not a good option for variable speed drive applications.

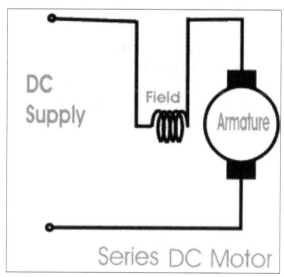

Figure-29. Series DC motor

3. **Shunt DC Motor** : In this type of DC motor, the field winding is connected in parallel to rotor winding so the voltage for both windings is same; refer to Figure-30. These motors offer great speed regulation due to the fact that the shunt field can be excited separately from the armature windings, which also offers simplified reversing controls.

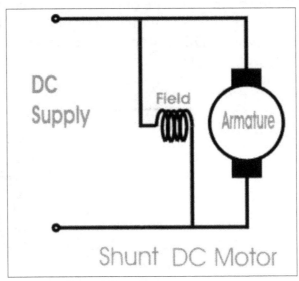

Figure-30. Shunt DC motor

4. **Compound DC Motor** : This type of DC motor uses combination of both shunt type and series type winding connections. Compound DC motors have good starting torque but may experience control problems in variable speed drive applications. There are two types of compound DC motors: cumulative and differential. When the shunt field flux assists the main field flux, produced by the main field connected in series to the armature winding then its called cumulative compound DC motor. In case of a differentially compounded self excited DC motor i.e. differential compound DC motor, the arrangement of shunt and series winding is such that the field flux produced by the shunt field winding diminishes the effect of flux by the main series field winding.

Brushless Electrical DC Motor

The brushless DC motors as name suggest do not have brushes to control rotation of motor. Instead a separate controller circuit is used to produce pulses of current for rotor or stator windings for controlling speed and torque of motor; refer to Figure-31.

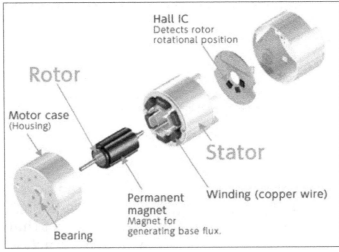

Figure-31. Brushless motor

Advantages of BLDC (BrushLess DC) motors :

- Brushless motors are more efficient as its velocity is determined by the frequency at which current is supplied, not the voltage.
- As brushes are absent, the mechanical energy loss due to friction is less which enhanced efficiency.
- BLDC motor can operate at high-speed under any condition.
- There is no sparking and much less noise during operation.
- More electromagnets could be used on the stator for more precise control.
- BLDC motors accelerate and decelerate easily as they are having low rotor inertia.
- It is high performance motor that provides large torque per cubic inch over a vast sped rang.
- BLDC motors do not have brushes which make it more reliable, high life expectancies, and maintenance free operation.
- There is no ionizing sparks from the commutator, and electromagnetic interference is also get reduced.
- Such motors cooled by conduction and no air flow are required for inside cooling.

Disadvantages of BLDC Motors:

- BLDC motor cost more than a brushed DC motor.
- The limited high power could be supplied to BLDC motor, otherwise, too much heat weakens the magnets and the insulation of winding may get damaged.

AC Motors

Alternating Current motors work on the principle of rotating magnetic field. The magnetic field of stator is made to rotate in circle and the electromagnetically charged rotor follows the rotation by attraction and repulsion due to polarity. There are mainly two types of AC motors : Induction Motor and Synchronous Motor.

Induction Motor

In an induction motor, there is no slip ring or DC excitation applied to rotor. The AC current in the stator induces a voltage across an air gap and into the rotor winding to produce rotor current and associated magnetic field. The stator and rotor magnetic fields then interact and cause the rotor to turn. The construction of induction motor is shown in Figure-32. There are two variations in induction motor: squirrel cage and wound rotor. In Squirrel cage induction motor, the rotor is constructed using a number of single bars short-circuited by end rings and arranged in a hamster-wheel or squirrel-cage configuration; refer to Figure-33. When voltage is applied to the stator winding, a rotating magnetic field is established. This rotating magnetic field causes a voltage to be induced in the rotor, which, because the rotor bars are essentially single-turn coils, causes currents to flow in the rotor bars. These rotor currents establish their own magnetic field, which interacts with the stator magnetic field to produce a torque

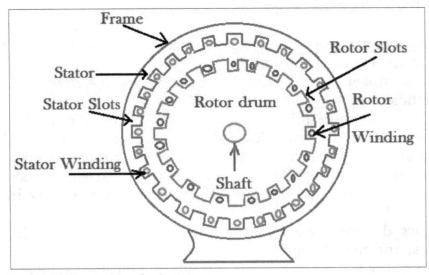

Figure-32. Construction of induction motor

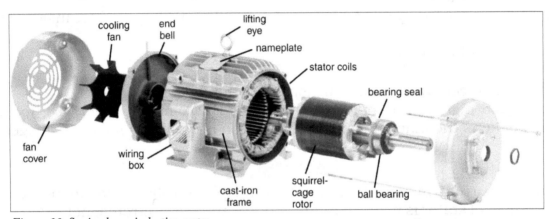

Figure-33. Squirrel cage induction motor

In wound rotor induction motor, also called slip-ring motor, wiring of rotor coils is terminated at slip rings; refer to Figure-34. A wound-rotor motor is used for constant-speed applications requiring a heavier starting torque than is obtainable with the squirrel-cage type. With a high-inertia load a standard cage induction motor may suffer rotor damage on starting due to the power dissipated by the rotor. With the wound rotor motor, the secondary resistors can be selected to provide the optimum torque curves and they can be sized to withstand the load energy without failure. Starting a high-inertia load with a standard cage motor would require between 400 and 550 percent start current for up to 60 seconds. Starting the same machine with a wound-rotor motor (slip-ring motor) would require around 200 percent current for around 20 seconds. For this reason, wound rotor types are frequently used instead of the squirrel-cage types in larger sizes.

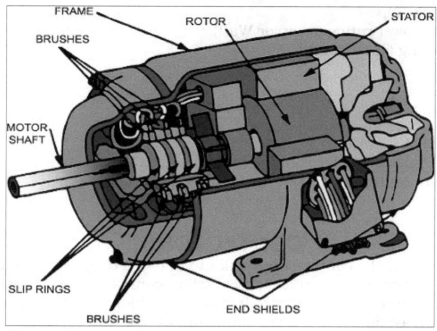

Figure-34. Wire wound induction motor

Synchronous Motor

In a synchronous ac motor, the stator magnetic field is produced by three phase power supply whereas the rotor magnetic field is produced by constant dc supply; refer to Figure-35. As you can see from the figure, the rotor behaves like permanent magnet with fixed polarity and follow the polarity force of stator. The rotor attracts towards the pole of the stator for the first half cycle of the supply and repulse for the second half cycle. Thus the rotor becomes pulsated only at one place. This type of motor are not self-starting and need an external method to bring rotor to synchronous speed.

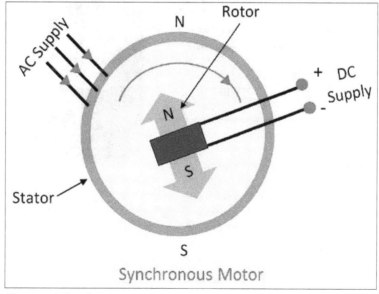

Figure-35. Synchronous motor

Selection of Motor

The electric motors are selected to satisfy the requirements of the machines on which they are to be installed. Following are some of the important factor the define the selection criteria of motor.

- Power Supply Type
- Mechanical power rating (in hp or Watt)
- Current Rating (in ampere)
- NEMA Code
- Efficiency
- Frame size
- Frequency
- Full-load RPM
- External requirements for running motor
- Temperature rating
- Duty cycle

CONTACTORS

The basic operation of a contactor is similar to that of a relay but contactor contacts can carry much more current than relays. Relays cannot be directly used in circuits where current exceeds 20 amperes. In such conditions contactors can be used. Contactors are available in a wide range of ratings and forms. Contactors are available up to the ampere rating of 12500A. Contactors cannot provide short circuit protection but can only make or break contacts when excited. When relatively lower voltage current is passed through coils of contactors, they cause high voltage supply contacts to close and when there is no current in the coils then contacts open; refer to Figure-36. There is no short-circuit protection in contactors so a separate circuit breaker or overload relay should be used with contactor in circuits.

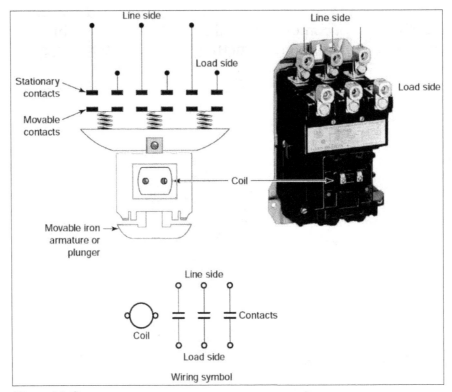

Figure-36. Contactor

Contactors are categorised based on the type of load (IEC utilisation categories - 60947) and current and power rating (NEMA size). Few important IEC utilisation categories are below:

AC-1: Non-inductive or slightly inductive and resistive heating type of loads

AC-2: Starting of slip ring induction motor
AC-3: Starting and switching off Squirrel-cage motors during running time
AC-15: Control of AC electromagnets.
AC-56b:- Switching of capacitor banks
DC–1: Non-inductive or slightly inductive and resistive heating type of loads
DC-2: Starting, inching and dynamic breaking of DC shunt motors
DC-3: Starting, inching and dynamic breaking of DC series motors
DC-13: Control of DC electromagnets

NEMA size:
NEMA size is based on the maximum continuous current and horse power rating of the induction motor controlled by the contactor. In NEMA standard contactors are designated as size 00,0,1,2,3,4,5,6,7,8,9. These numbers represent current ratings; refer to Figure-37.

60 Hz AC contactor NEMA ratings 600 volts max		DC contactor NEMA ratings 600 volts max	
NEMA size	Continuous amps	NEMA size	Continuous amps
00	9	1	25
0	18	2	50
1	27	3	100
2	45	4	150
3	90	5	300
4	135	6	600
5	270	7	900
6	540	8	1350
7	810	9	2500
8	1215		
9	2250		

Figure-37. NEMA ratings for contactors

Solid State Contactor

In solid state contactor; refer to Figure-38, electronic switching is used to open or close contacts of main line. In this type of contactor, there is no magnetic coil. Instead, semiconductors are used to perform contactor work. The most common high-power switching semiconductor used in solid-state contactors is the silicon controlled rectifier (SCR). An SCR is a three-terminal semiconductor device (anode, cathode, and gate) that acts like the power contact of a magnetic contactor. A gate signal, instead of an electromagnetic coil, is used to turn the device on, allowing current to pass from cathode to anode. Since there is not moving part so they are faster in switching as compared to magnetic contactors. The only drawback with solid state contactors is that when they fail, they fail causing short circuit instead of open circuit. So, they can become dangerous if not made in good quality.

Figure-38. Solid state contactor

MOTOR STARTERS

Motor starter is a combination of contactors, disconnectors, and overload protection circuits. When starting a motor, the circuit has to be upto 20 times more current than regular current required to run motor at normal speed. So, the system should have an overload circuit that counts for this phase. Once the motor is running on normal speed then it should protect the motor in high current or short-circuit mishaps. Figure-39 shows a typical motor starter construction.

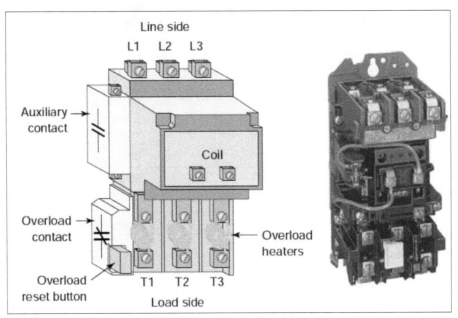

Figure-39. Motor starter

There are mainly four types of motor starters: Across-The-Line, the Reversing Starter, the Multispeed Starter, and the Reduced Voltage Starter.

Across-the-line or Full Voltage Non-Reversing (FVNR) is the most commonly used general purpose starter. This starter connects the incoming power directly to the motor. It can be used in any application where the motor runs in only one direction, at only one speed, and starting the motor directly across the line does not create any "dips" in the power supply.

The **Reversing Starter or Full Voltage Reversing (FVR)** reverses a motor by reversing any two leads to the motor. This is accomplished with two contactors and one overload relay. One contactor is for the forward direction and the other is for reverse. It has both mechanically and electrically interlocked sets of contactors.

The **Multispeed Starter** is designed to be operated at constant frequency and voltage. There are two ways to change the speed of an AC motor: Vary the frequency of the current applied to the motor or use a motor with windings that may be reconnected to form different number of poles. The multispeed starter uses the latter option to change speed.

Reduced Voltage Starter (RVS) is used in applications that typically involve large horsepower motors. The two main reasons to use a reduced voltage starter are to reduce the inrush current and to limit the torque output and mechanical stress on the load. Power companies often won't allow this sudden rise in power demand. The reduced voltage starter addresses this inrush problem by allowing the motor to get up to speed in smaller steps, drawing smaller increments of current. This starter is not a speed controller. It reduces the shock transmitted to the load only upon start-up.

FUSES

Fuses are the most simple devices used to protect circuit from current overload. Fuses are sacrificial devices which get blown when there is too much current through them. In a fuse, wire of specific size is installed between two terminals. This wire has fixed rating in amperes upto which it can pass the current. When current of higher level is passed, it gets melt and disconnects the circuit. You can find fuses in almost every electrical appliance. Fuses are divided into two categories, Low Voltage Fuse (LV Fuse) and High Voltage Fuse (HV Fuse). The low voltage fuses are good upto 230 V and beyond that you should use HV fuses. The low voltage fuses are further divided into four categories rewire-able, cartridge, striker and switch fuses; refer to Figure-40.

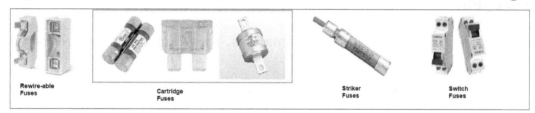

Figure-40. Low Voltage fuses

The high voltage fuses are used to protect transformers and power supplies. There are mainly three types of high voltage fuses; Cartridge Type HRC Fuse, Liquid Type HRC Fuse, and Expulsion Type HRC Fuse; refer to Figure-41.

Selection of a fuse depends on the normal current rating of connected components that you want to protect.

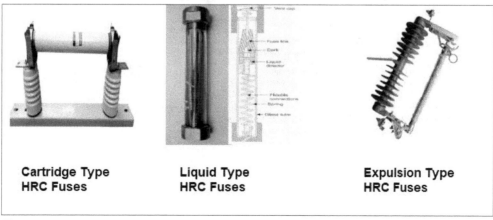

Figure-41. High voltage Fuses

CIRCUIT BREAKERS

Circuit Breaker as the name suggests is used to break the circuit when current is above the specified limits. The functioning of circuit breaker is similar to fuse as it also protects circuit from high current. The difference is that a circuit breaker changes from closed to open position when current is higher than its limits (It does not become sacrificial lamb). When fault has been identified and cured then you use the same circuit breaker again. The circuit breakers are available in three categories: Standard, GFCI and AFCI.

The standard circuit breakers are available in two configurations single pole and double-pole. The single pole circuit breaker can handle 120V supply with 15 amp to 30 amp current where as double pole circuit breaker can handle 230V supply with 15 to 200 amps of current. The standard circuit breakers are good for normal appliances.

The GFCI (Ground Fault Circuit Interrupter) Circuit Breakers are used when a line to ground fault can occur causing life threatening situations to consumer like in wet areas, workshops, and areas where power tools are used.

The AFCI (Arc Fault Circuit Interrupter) Circuit Breakers cut the supply before arc is formed in the circuit and protect from possible fire hazard. Even a small surge of current can cause the AFCI circuit breaker to trip.

Now a days, you can find circuit breakers that combine the benefits of all the three types in one unit.

Depending the applications of circuit breakers, they can be classified as shown in Figure-42.

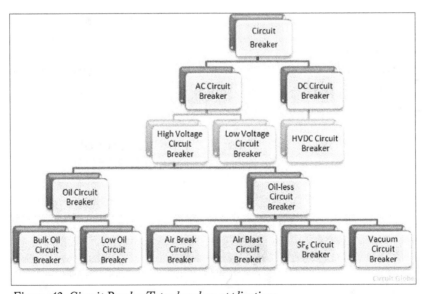

Figure-42. Circuit Breaker Types based on applications

DISCONNECTORS

Disconnectors are used to completely de-energize the circuits for maintenance work. A disconnector should also ensure that no accidental connection is formed when circuit is open. Disconnectors are generally used in high voltage transmission lines carrying current at 32 A to 2000 A with voltage ranging from 36 kV to 100kV. You can find disconnectors for low voltage circuits also but contactors and circuit breakers are good alternatives in such cases. The disconnectors used in low voltage applications are called switch disconnectors or safety switches; refer to Figure-43.

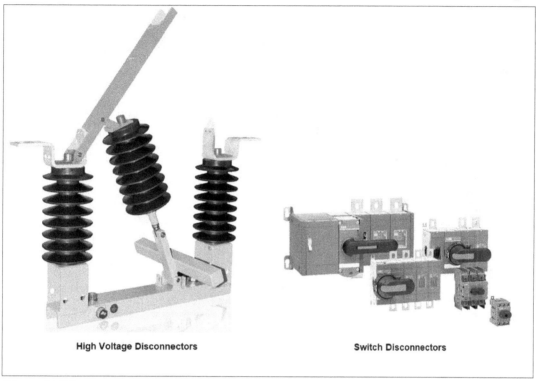

High Voltage Disconnectors   **Switch Disconnectors**

Figure-43. Disconnectors

TERMINALS AND CONNECTORS

Electric Terminal are used to connects wires and systems with low resistance. In every well designed system, the wires should end with terminals for connections. A connector has fixed number of terminals in which wires are fixed. Most of the time, you will find connectors in pair: Male connector and Female connector which fit in each other to form connection. There are various types of terminals are shown in Figure-44. A 9 pin male and female connector pair is shown in Figure-45.

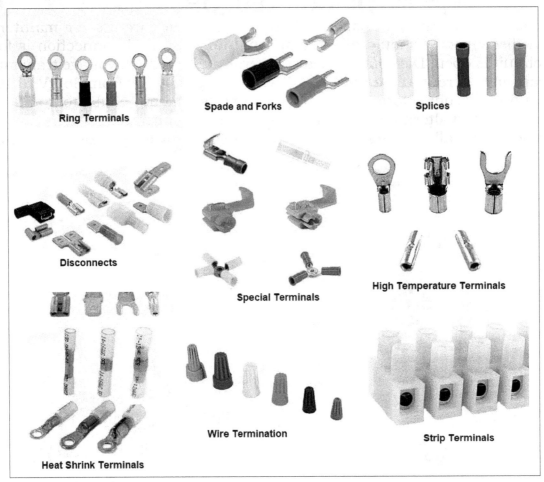

Figure-44. Terminals

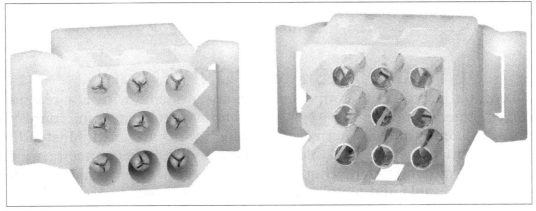

Figure-45. Electrical connectors

Index